NATURAL DISASTER RESEARCH, PREDICTION AND MITIGATION

THE COMMUNITY DISASTER LOAN PROGRAM

HISTORY, ANALYSES AND CONSIDERATIONS

Natural Disaster Research, Prediction and Mitigation

Additional books in this series can be found on Nova's website
under the Series tab.

Additional E-books in this series can be found on Nova's website
under the E-book tab.

Government Procedures and Operations

Additional books in this series can be found on Nova's website
under the Series tab.

Additional E-books in this series can be found on Nova's website
under the E-book tab.

THE COMMUNITY DISASTER LOAN PROGRAM

HISTORY, ANALYSES AND CONSIDERATIONS

FELIX P. CARDANO
EDITOR

New York

For permission to use material from this book please contact us:
Telephone 631-231-7269; Fax 631-231-8175
Web Site: http://www.novapublishers.com

NOTICE TO THE READER

Library of Congress Cataloging-in-Publication Data

ISBN: 978-1-62417-643-2

Published by Nova Science Publishers, Inc. ✝ New York

CONTENTS

PREFACE

The core purpose of the Community Disaster Loan (CDL) program is to provide financial assistance to local governments that are having difficulty providing government services because of a loss in tax or other revenue following a disaster. The program assists local governments by offering federal loans to compensate for this temporary or permanent loss in local revenue. The CDL program is managed by the Federal Emergency Management Agency (FEMA). This book explores the history, analysis and issues for Congress of FEMA's Community Disaster Loan Program with a focus on regulations, funding and eligibility criteria.

Chapter 1 – The core purpose of the Community Disaster Loan (CDL) program is to provide financial assistance to local governments that are having difficulty providing government services because of a loss in tax or other revenue following a disaster. The program assists local governments by offering federal loans to compensate for this temporary or permanent loss in local revenue. The CDL program is managed by the Federal Emergency Management Agency (FEMA). First authorized in the Disaster Relief Act of 1974 (P.L. 93-288), the Community Disaster Loan program is currently codified in Section 417 of the Robert T. Stafford Disaster Relief and Emergency Assistance Act (42 U.S.C. §5184, as amended). The program is funded through the Disaster Assistance Direct Loan Program account, rather than the Disaster Relief Fund (DRF) that funds the majority of other Stafford Act programs. In sum, 249 loans were issued to 200 local governments under 26 different disaster declarations from 1974 to 2010. An approximate total of $1,615 million in principal was offered to these governments in loans, of which roughly $1,326 million was borrowed by the governments. Through the program, FEMA may also cancel the repayment of the loans if certain financial conditions prevailed after the three fiscal years following the disaster.

Through its cancellation authority, FEMA has forgiven approximately $896 million of the $1,326 million in principal advanced to local governments since program inception. This report compares and analyzes three different categories of loans issued in different time periods in the program's history: "traditional" loans issued between 1974-2005, in 2007, and between 2009 and June 2012 (TCDLs); "special" (SCDLs) loans issued in 2005-2006 following Hurricanes Katrina and Rita; and loans issued under unique provisions in 2008 (2008 CDLs). As authorized by Congress and administered by FEMA, the SCDL and 2008 loan categories had different provisions than traditional loans to guide the eligibility of local governments and dollar size of the loans. SCDLs also had unique provisions that slightly altered the purpose of the loans, lowered the interest rate charged on the loans, and clarified the cancellation procedures for the loans. In the original legislation authorizing and appropriating the SCDLs, repayment of the loans was not allowed to be cancelled by FEMA. However, Congress later amended the law to allow cancellation for SCDLs. Some controversy has arisen over FEMA's administration of the cancellation authority for these special loans, with many suggesting that FEMA has not cancelled the appropriate amount of loan balances. Table 8 and Table 9 provide several measures for comparing the cancellation rates of TCDLs to SCDLs. In summary, TCDLs had a lower percentage of loans fully cancelled or with some level of cancellation than SCDLs (33.9% and 46.4% versus 50.0% and 59.7%, respectively). On average, TCDLs also had lower dollar amounts of principal forgiven per loan than SCDLs (38.9% versus 54.1%). However, as a function of total dollar amount of principal cancelled in each loan category, TCDLs had a much higher cancellation rate than SCDLs (97.2% versus 68.9%). As reported out of the Senate Appropriations Committee, the Department of Homeland Security Appropriations Act, 2013 (S. 3216) includes a provision that would alter existing cancellation procedures for SCDLs. The provision may require FEMA to reopen the review process for 71 special loans, issued to 54 local governments, that have not been fully cancelled in existing procedures. If passed into law, this provision may result in higher rates of cancellation for the SCDLs. Congress may consider changes to the overall CDL program in the future. Options could include altering future authorization and appropriations for the program in favor of more tailored disaster assistance programs, or converting the loan assistance into a grant program. There are also options for amending the program less significantly, including changing the way loan funds may be used by local governments, changing the total dollar size of the loans, and altering how the cancellation authority can be applied by FEMA.

Chapter 2 – On August 2, 2011, the President signed into law the Budget Control Act of 2011 (BCA, P.L. 112- 25), which included a number of budget-controlling mechanisms. As part of the legislation, caps were placed on discretionary spending for the next ten years, beginning with FY2012. If these caps are exceeded, an automatic rescission—known as sequestration—takes place across most discretionary budget accounts to reduce the effective level of spending to the level of the cap. Additionally, special accommodations were made in the BCA to address the unpredictable nature of the need for disaster assistance while attempting to impose discipline on the amount spent by the federal government on disasters. The first section of this report addresses the traditional funding for major disaster declarations, both through annual requested amounts and through supplemental appropriations to meet greater, unanticipated costs. This section also explains the workings of the President's Disaster Relief Fund, a "no-year" fund that finances spending under the Robert T. Stafford Disaster Relief and Emergency Assistance Act (P.L. 93-288). Next, this report provides a basic overview of how disaster assistance is appropriated, what factors affect how much the federal government spends on disasters, how disaster relief is impacted by the BCA, and what the policy implications are for disaster assistance going forward under the constraints of the BCA. Included in this review are discussions of disaster spending, how it is calculated under the BCA, and potential costs that may be excluded under that calculation. The report also touches on the increasing number of disaster declarations and both the possible causes and likely cost implications of a greater number of declarations.

Chapter 3 – P.L. 112–25, the Budget Control Act of 2011 (BCA), amended the Balanced Budget and Emergency Deficit Control Act of 1985 (BBEDCA), by reinstating the discretionary spending limits that had expired after 2002 and by creating ad-justments to those limits. Section 251(b)(2)(D)(ii) of BBEDCA requires the Office of Management and Budget (OMB) to submit a report to the Committees on Appropriations and the Budget in the U.S. House of Representatives and the Senate within 30 days of the enactment of the BCA on the average funding provided for disaster relief over the previous 10 years (excluding the highest and lowest years). In addition, section 254(e) of BBEDCA, as amended by section 103(2) of the BCA, requires OMB to include in its Sequestration Update Report a "preview estimate of the adjustment for disaster funding for the upcoming fiscal year," in this case FY 2012.

The average funding provided for disaster relief over the previous 10 years (excluding the highest and lowest years) is $11.5 billion for fiscal year 2011 and $11.3 billion for fiscal year 2012. There is an inadequate basis to estimate the final adjustment for fiscal year 2012 because of the current status of the appropriations process and the newness of the disaster relief adjustment in BBEDCA. The amount of $11.3 billion represents a ceiling on the potential disaster adjustment, but the actual level of the adjustment will be decided through the fiscal year 2012 appropriations process. Applying OMB's methodology for determining prior year levels of funding for disaster relief to the specific language in the President's fiscal year 2012 budget request and other known disaster relief needs as identified in prior law would support a disaster relief adjustment equal to $5.2 billion. In addition, there is no question that additional funds will be required to assist with the response to Hurricane Irene, on top of the $5.2 billion. As the Administration gathers data about the extent of the damage caused by Hurricane Irene, the needed amounts will promptly be determined and provided to the Congress during its appropriations process this month.

In: The Community Disaster Loan Program   ISBN: 978-1-62417-643-2
Editor: Felix P. Cardano   © 2013 Nova Science Publishers, Inc.

Chapter 1

FEMA's Community Disaster Loan Program: History, Analysis, and Issues for Congress[*]

Jared T. Brown

SUMMARY

The core purpose of the Community Disaster Loan (CDL) program is to provide financial assistance to local governments that are having difficulty providing government services because of a loss in tax or other revenue following a disaster. The program assists local governments by offering federal loans to compensate for this temporary or permanent loss in local revenue. The CDL program is managed by the Federal Emergency Management Agency (FEMA). First authorized in the Disaster Relief Act of 1974 (P.L. 93-288), the Community Disaster Loan program is currently codified in Section 417 of the Robert T. Stafford Disaster Relief and Emergency Assistance Act (42 U.S.C. §5184, as amended). The program is funded through the Disaster Assistance Direct Loan Program account, rather than the Disaster Relief Fund (DRF) that funds the majority of other Stafford Act programs. In sum, 249 loans were issued to 200 local governments under 26 different disaster declarations from 1974 to 2010. An approximate total of

---

[*] This is an edited, reformatted and augmented version of the Congressional Research Service Publication, CRS Report for Congress R42527, dated July 3, 2012.

$1,615 million in principal was offered to these governments in loans, of which roughly $1,326 million was borrowed by the governments. Through the program, FEMA may also cancel the repayment of the loans if certain financial conditions prevailed after the three fiscal years following the disaster. Through its cancellation authority, FEMA has forgiven approximately $896 million of the $1,326 million in principal advanced to local governments since program inception.

This report compares and analyzes three different categories of loans issued in different time periods in the program's history: "traditional" loans issued between 1974-2005, in 2007, and between 2009 and June 2012 (TCDLs); "special" (SCDLs) loans issued in 2005-2006 following Hurricanes Katrina and Rita; and loans issued under unique provisions in 2008 (2008 CDLs). As authorized by Congress and administered by FEMA, the SCDL and 2008 loan categories had different provisions than traditional loans to guide the eligibility of local governments and dollar size of the loans. SCDLs also had unique provisions that slightly altered the purpose of the loans, lowered the interest rate charged on the loans, and clarified the cancellation procedures for the loans.

In the original legislation authorizing and appropriating the SCDLs, repayment of the loans was not allowed to be cancelled by FEMA. However, Congress later amended the law to allow cancellation for SCDLs. Some controversy has arisen over FEMA's administration of the cancellation authority for these special loans, with many suggesting that FEMA has not cancelled the appropriate amount of loan balances. **Table 8** and **Table 9** provide several measures for comparing the cancellation rates of TCDLs to SCDLs. In summary, TCDLs had a lower percentage of loans fully cancelled or with some level of cancellation than SCDLs (33.9% and 46.4% versus 50.0% and 59.7%, respectively). On average, TCDLs also had lower dollar amounts of principal forgiven per loan than SCDLs (38.9% versus 54.1%). However, as a function of total dollar amount of principal cancelled in each loan category, TCDLs had a much higher cancellation rate than SCDLs (97.2% versus 68.9%). As reported out of the Senate Appropriations Committee, the Department of Homeland Security Appropriations Act, 2013 (S. 3216) includes a provision that would alter existing cancellation procedures for SCDLs. The provision may require FEMA to reopen the review process for 71 special loans, issued to 54 local governments, that have not been fully cancelled in existing procedures. If passed into law, this provision may result in higher rates of cancellation for the SCDLs.

Congress may consider changes to the overall CDL program in the future. Options could include altering future authorization and appropriations for the program in favor of more tailored disaster assistance programs, or converting the loan assistance into a grant program. There are also options for amending the program less significantly, including changing the way loan funds may be used by local governments, changing the total dollar size of the loans, and altering how the cancellation authority can be applied by FEMA.

OVERVIEW

Following a major disaster, the financial capacity of a local government may be severely undermined by a decrease in local revenues. The reduction in tax or other revenue can limit the local government's ability to maintain public services or afford many extraordinary but necessary expenditures. Revenue shortfalls can also impact the size of a local government and the jobs of government employees, as exemplified by the decrease in the size of New Orleans city government since Hurricane Katrina.[1] Revenue loss frequently occurs when significant portions of the population are displaced for extended periods of time, or key sources of economic activity, like tourism, are heavily disrupted by a disaster. In addition, many local governments are restricted by state or local laws, constitutions, or codes of practice from borrowing to fund operational expenses.[2] Often with few options available to make up for lost revenue, local governments may need to reappropriate funds from other portions of their budget or scale back their operations. The shortage of revenues, and the resulting limitation on financial capacity, has been cited as one of the most significant and consistent hurdles to long-term disaster recovery.[3]

The Community Disaster Loan (CDL) Program, managed by the Federal Emergency Management Agency (FEMA), provides loan assistance to local governments to help them overcome a loss in revenues. Though there are many disaster assistance programs available to communities, the CDL program is the only program that specifically provides assistance to local governments to help compensate for revenue shortfalls. The core purpose of these community disasters loans, as detailed in the original Senate committee report authorizing the program, is "to permit the local governments to continue to provide municipal services, such as the protection of public health and safety and the operation of the public school system."[4] Consistent with this stated purpose, the local governments that are eligible for loans include entities such as special districts and school boards that provide a wide array of local government services.[5]

The CDL program was first authorized by the Disaster Relief Act of 1974, which was later renamed the Robert T. Stafford Disaster Relief and Emergency Assistance Act (Stafford Act).[6] As codified in Section 417 of the Stafford Act, the program allows the President to

> make loans to any local government which may suffer a substantial loss of tax and other revenues as a result of a major disaster, and has demonstrated a need for financial assistance in order to perform its governmental functions.[7]

As with most other authorities in the Stafford Act, this authority has been delegated from the President to the Administrator of FEMA.[8]

After severe disasters, the revenue base of a local government may take years to fully recover, if ever. Some key sources of revenue may never return to pre-disaster levels, such as property taxes from areas severely damaged by a disaster. To account for the local government's continuing need for financial assistance in these circumstances, FEMA also has the authority to cancel the repayment on all or part of the loan. FEMA may cancel a loan up to the amount that the local government's tax and other revenues are insufficient to meet its cumulative operating budget over a period of three full fiscal years following a major disaster, in addition to any unreimbursed disaster-related expenses made in that time period.[9]

The fundamental purpose of the program has been relatively unchanged since inception. However, some statutory and regulatory provisions have varied through the history of the program, notably those that govern elements such as loan eligibility and the size of the loans. The loan program that began in 1974 was essentially unaltered in statute until 2000, when the 106th Congress revised the core polices of the program for the first time by placing a $5 million cap on the size of the loans. Loans administered before 2005, and recent loans not governed by the exemptions discussed immediately below, are known as "traditional" community disaster loans (TCDLs).

Following Hurricane Katrina and the 2005 hurricane season, the 109th and 110th Congresses created unique statutory guidelines for the loan program applicable exclusively to local governments impacted by those disasters. Collectively, these loans are referred to as "special" community disaster loans (SCDLs). FEMA promulgated regulations to govern the implementation of SCDLs that were very similar to those for the TCDLs, but with several notable distinctions.[10] These new statutory and regulatory guidelines altered some of the eligibility, size, and forgiveness criteria applicable to the SCDLs.

A similar pattern of special legislation was again passed in the 110th and 111th Congresses for community disaster loans applicable to local governments impacted by disasters in the calendar year 2008. These loans also had some different provisions for eligibility and size, as will be discussed, and will be referred to as the 2008 CDLs for the purposes of distinguishing them in this report.[11]

The majority of the statutory and regulatory provisions of the Community Disaster Loan Program are consistent across each set of loans. While this report will explain the consistent provisions of the loan program, it is intended also to highlight the unique provisions through a comparative analysis of traditional CDLs, special CDLs, and 2008 CDLs. In particular, since the issue of loan forgiveness has been a flash point of policy debate throughout the program's history, this report also highlights the statutory and regulatory provisions for loan forgiveness and discusses some of the debate on the issue. In doing so, the report offers some empirical analysis of data from each set of loans to identify the distribution and forgiveness of CDLs by the types of governments receiving a loan.[12] It is important to note that this report is not intended to comprehensively audit FEMA's administration or forgiveness of any particular loan or set of loans.

The CDL Program may be of interest to Congress because of the ongoing debate over the cost and amount of disaster assistance provided to communities.[13] Moreover, as will be discussed in this report, recent program management decisions by FEMA have drawn the attention and criticism of some in the general public and of Members of Congress. Congress may be interested in expanding, eliminating, or amending the CDL program before additional loans are issued.

HISTORY OF USE, REGULATIONS, AND FUNDING

Frequency of Program Use

The historical use of the program is summarized in **Table 1** and **Table 2**. An approximate total of $1,615 million in principal was offered to local governments through the program, with roughly $1,326 million borrowed. FEMA has cancelled approximately of $896 million of the $1,326 million of principal advanced to the local governments since program inception.

Historically, the Community Disaster Loan Program has been used infrequently by local governments, relative to other disaster assistance programs authorized by the Stafford Act.[14] In sum, 249 loans were issued to 200 local governments under 26 different disaster declarations from 1974 to 2010. In that same period, there were 1,542 disaster declarations leading to an incalculable number of local governments conceivably eligible for loans. From August 1976 through September 30, 2005, a period of 29 years, FEMA approved 64 traditional loans for different local governments related to 21 separately declared disasters. Following Hurricanes Katrina and Rita in 2005, FEMA issued 157 special loans to 109 local governments. In response to the 2008 disasters, FEMA issued 24 loans to 23 different local governments. Program use is generally confined to a large number of loans being issued following specific disasters. For instance, FEMA issued 28 loans after floods in Illinois in 1993, 157 loans after Hurricanes Katrina and Rita, and 14 loans after Hurricane Ike in 2008. But no loans were issued after the 9/11 terrorist attacks, after any earthquakes (including the Northridge earthquake in 1994), or after several major hurricanes like Floyd in 1999, Isabel in 2003, or the series of four Florida hurricanes in 2004. No community disaster loans were issued from FY1999 through FY2005.

The limited use of the program may be attributable to a number of reasons, including, but not limited to:

- A general lack of awareness of the program among local governments impacted by disasters, or limited advertisement of the program by FEMA to those governments;
- The eligibility provisions of the loan program may exclude many potential applicants;
- The size, interest rate, or others terms of the loan may be unattractive to local governments;
- The procedures for applying and managing the loan may be considered too cumbersome or intrusive by local or state governments following a disaster;
- There is relatively little need for the program overall, except in particular disaster situations, in comparison to other programs. This may be caused, in part, by circumstantial immediate increases in local revenues produced by the often high level of economic activity involved with the disaster response and recovery efforts.

Table 1. Summary of the Approved Number of Community Disaster Loans

Category of CDL	Number of Approved Loans	Number of Recipient Local Governments	Number of Loans Borrowed[a]	Number of Borrowing Local Governments[b]	Related Number of Declared Disasters	Related Time Period
Traditional CDLs	68	68	57	57	21	1976-2005, 2007, 2009 to present
Special CDLs	157	109	129	92	2[c]	2005-2006
2008 CDLs	24	23	20	19	3	2008
Total	**249**	**200**	**206**	**168**	**26**	**1976 - June 15, 2012**

Source: Calculations and categorization by CRS. Raw data provided by FEMA staff, and were current as of June 15, 2012.

Notes:

a. This column includes only loans where the recipient local governments chose to borrow some of the approved amount of principal.

b. This column counts only the number of recipient local governments that chose to borrow some of the approved amount of principal. Some local governments borrowed more than one loan, as allowable under the "Special" and 2008 loan categories.

c. These loans were technically issued under two different disaster declarations following Hurricane Katrina—the separate declarations for Louisiana and Mississippi.

An accurate accounting of the number and type of potential applicants to the program, and their reasons for not applying, is not available. Therefore, it is difficult to assess the relatively limited demand for the CDL program.

Regulations

It is beyond the scope of this report to fully discuss each iteration of the regulations governing the CDL program since 1974. However, a summary of rulemakings is provided in **Table 3**. The core policies of the program were initially established in a program manual, and then promulgated in a rulemaking in 1979. The CDL program rules did not undergo dramatic policy change between 1979 and 2005, though considerable clarification and specificity has been added over various rulemakings, most prominently in the

1988 rulemaking. As warranted, notable changes in regulations are remarked upon further throughout the report.

Table 2. Summary of the Approved and Borrowed Dollar Amounts of Community Disaster Loans, in Millions of Dollars

Category of CDL	Amount of Approved Principal	Amount of Principal Borrowed	Amount of Accrued Interest on Loans	Amount of Principal Cancelled	Amount of Interest Cancelled
Traditional CDLs	$284	$236	$108	$180[a]	$79[a]
Special CDLs	$1,271	$1,041	$131	$717	$79
2008 CDLs	$61	$50	$3	N/A[b]	N/A[b]
Total	**$1,615**	**$1,326**	**$242**	**$896**	**$158**

Source: Calculations and categorization by CRS. Raw data provided by FEMA staff, and were current as of June 15, 2012.

Notes: Dollars listed in millions. Dollar figures have been rounded to nearest million. Totals may not add due to rounding error. SCDLs in appeal (17) are included by their cancellation status as of June 15, 2012. Four of the appealed loans have already been partially cancelled, and the remaining 13 have had no amount cancelled.

a. These figures exclude a U.S. Virgin Island loan that was cancelled in part by legislative action was excluded from calculations. See discussion in "Summary of Traditional Loan Cancellation."

b. These loans have not yet been reviewed for cancellation by FEMA; see later discussion in "Timeline for 2008 Community Disaster Loan ."

Funding

Unlike most other Stafford Act programs, the Community Disaster Loan program is not funded through the Disaster Relief Fund. The program is instead funded through the Disaster Assistance Direct Loan Program (DADLP) account.[15] The account also funds activities under Section 319 of the Stafford Act, which provides advances or loans for the portion of assistance applicants are responsible for under the different cost-sharing provisions of the Stafford Act (commonly referred to as the applicant/state "match" for assistance).[16] Generally, funds have been annually appropriated to the DALDP account for the purposes of Section 319 of the Stafford Act. However, funds for the purposes of the Community Disaster Loan program have often been appropriated through emergency supplemental appropriation bills in response to a particular set of disasters. Like the Disaster Relief Fund, funds appropriated to the DALDP have traditionally been treated as "no year" funds and were available until expended. **Table 4** provides a list of the appropriations for the program by since 1990.

Table 3. Summary of Rulemakings on the Community Disaster Loan Program

Federal Register Citation	Summary of Action
Department of Housing and Urban Development, "Federal Disaster Assistance; Community Disaster Loans," 44 *Federal Register* 47105 - 47109, August 10, 1979.	First proposed rule for the CDL program. Prior to this publication, policy was established in an executive manual for disaster assistance.
Federal Emergency Management Agency, "Disaster Assistance: Community Disaster Loans," 44 *Federal Register* 71790, December 11, 1979.	Same as proposed rule with one minor definition change. Rule established the core policies for loan eligibility, application, and cancellation used throughout the program's history.
Federal Emergency Management Agency, "Disaster Assistance; Community Disaster Loans," 52 *Federal Register* 15348, April 28, 1987.	Proposed rulemaking to "improve the program and clarity of the regulations."
Federal Register Citation	**Summary of Action**
Federal Emergency Management Agency, "Disaster Assistance; Community Disaster Loans," 53 *Federal Register* 12681, April 18, 1988.	Final rule implementing a series of clarifications and improvements to the regulations. Major items included: requiring a state to co-sign a loan or the local government to provide collateral for the loan; clarified the types of information required from the local government when applying and submitting for cancellation; and allowed repayment of loans to extend beyond 10 years under certain extenuating circumstances.
Federal Emergency Management Agency, "Disaster Assistance," 54 *Federal Register* 11610, March 21, 1989.	Part of a much larger proposed rule for overall disaster assistance. Only administrative changes proposed for the CDL program.
Federal Emergency Management Agency, "Disaster Assistance," 55 *Federal Register* 2297, January 23, 1990.	Final rule renumbered sections of the CDL regulations in order to accommodate an expansion of the overall disaster assistance regulations. No policy impact for CDL program.
Department of Homeland Security, "Special Community Disaster Loans Program," 70 *Federal Register* 60443, October 18, 2005.	Interim rule for the special community disaster loan program. See discussion in "Background on Special Community Disaster Loan Regulations."
Department of Homeland Security, "Special Community Disaster Loans Program," 74 *Federal Register* 15228, April 3, 2009.	Proposed rule for the special community disaster loan program. See discussion in "Background on Special Community Disaster Loan Regulations."
Department of Homeland Security, "Special Community Disaster Loans Program," 75 *Federal Register* 2800, January 19, 2010.	Final rule for the special community disaster loan program. See discussion in "Background on Special Community Disaster Loan Regulations."

Source: CRS analysis of rulemakings in the Federal Register.

Table 4. Appropriations to the DALDP account for Community Disaster Loans, 1990-2012

Fiscal Year	Public Law and U.S. Statutes Citation	Appropriated Amount, in Millions
1992	P.L. 102-389; 106 Stat. 1604-1605	$50.0 and $200.0[a]
1997	P.L. 105-18; 111 Stat. 158	$20.0
2006	P.L. 109-88; 119 Stat. 2061	$750.0
2006	P.L. 109-234; 120 Stat. 459-450	$279.8
2008	P.L. 110-329; 122 Stat. 392	$98.2

Source: CRS analysis of enacted appropriation bills.

Notes: Appropriations to other accounts that may have funded community disaster loans through reprogramming requests not noted in statute are not captured in this table. This table does not catalogue appropriations to the DALDP for administrative expenses. Dollar figures have been rounded to nearest tenth of a million.

a. This law made two separate appropriations for CDLs. $50 million was appropriated until expended by amending a previous law, the Dire Emergency Supplemental Appropriations Act,1992 (P.L. 102-302), through a transfer of funds from the DRF. The second appropriation, for $200 million, was appropriated but only to be used until the end of FY1993.

The CDL program is subject to the Federal Credit Reform Act of 1990, as amended (P.L. 101508, FCRA).[17] The FCRA changed the accounting method for measuring the cost of federal direct loans and loan guarantees, starting in FY1992. Under the FCRA, discretionary programs providing new direct loan obligations or new loan guarantee commitments require appropriations of budget authority equal to the loans' estimated subsidy costs. Furthermore, the appropriations bill must include an estimate for the dollar amount of the new direct loan obligations that are supported by the subsidy budget authority appropriated to the agency for its credit program. The subsidy rate for any loan program is calculated by CBO in each appropriation. Therefore, appropriations to the DADLP account for the purposes of the CDL program have varied in the total dollar amount of loans that can be issued per appropriated dollar.[18] For example, in P.L. 109-88, a congressional appropriation to the DALDP account of $750 million in budget authority actually supported the issuance of $1 billion in loans, at a calculated subsidy rate of 75%. However, in P.L. 110-329 the DADLP account was appropriated $98.15 million for the CDL program, to subsidize gross obligations for the principal amount of direct loans not to exceed $100 million, at a calculated subsidy rate of 98.15%.[19]

The Office of Management and Budget also produces subsidy rate estimates for each program in the annual budget report. However, since loans were last issued by FEMA in 2009, the latest available rate estimate for the CDL program was in FY2009, at a rate of 93.95%.[20] This rate is very high in relation to other loan programs because of the high cancellation rate of the program (discussed in full later).

TRADITIONAL COMMUNITY DISASTER LOANS

This section of the report describes the law and regulations that governed all traditional community disaster loans (TCDLs) issued from April 1988 to October 2005, and all loans since 2005 not covered by the unique provisions of the SCDLs or the 2008 loans discussed later. It also describes important statutory and regulatory elements that are applicable for all categories of loans in the program. By default, these traditional rules will continue to govern all future community disaster loans unless otherwise mandated by Congress or until FEMA revises the regulations. For the purposes of calculations and categorization, however, this report classifies all loans made through the CDL program from inception to 2005 as "traditional," since the loans were guided by relatively the same regulations. Of note, there were 64 traditional loans issued between 1976 and 1998, with no loans issued between 1998 and 2005. Four loans governed by the traditional regulations were issued in response to the severe tornadoes that struck Greensburg, KS, and surrounding areas in 2007.

Conditions for Eligibility, Loan Size, and Administration

Eligibility Criteria

To be eligible for a loan, local governments must meet three basic requirements provided in Section 417 of the Stafford Act. First, local governments must have suffered losses "as a result of a major disaster." Second, a local government should have "suffer[ed] a substantial loss of tax and other revenues." Third, there "has to be a demonstrated need for financial assistance" for the loan to be issued.[21] FEMA's interpretation of these three statutory conditions are formulated in the eligibility criteria for traditional loans outlined in 44 C.F.R. §206.363.

Consistent with the first requirement, FEMA regulations restrict eligibility to only those local governments in the designated area of a presidential major disaster declaration.[22] Local government is broadly defined in the Stafford Act,[23] and therefore loans have been granted to governmental bodies ranging from general purpose municipal city governments, tribes, hospitals, school boards, sheriff departments, water and sewage authorities, to transportation districts. FEMA has previously determined that this definition also treats each U.S. territory as a single eligible municipal taxing authority, which has allowed the Virgin Islands and American Samoa to submit consolidated applications.[24] Local governments are ineligible, however, if they are legally barred from incurring federal debt, or debt to fund operating expenses, either by state or local laws. **Table 5** displays the distribution of community disaster loans by type of local government entity.

Table 5. Number of Eligible Governments Receiving Loans

	Traditional Loans	Special Loans	2008 Loans	Total
Territory[a]	3	0	0	3
Municipal Governments[b]	42	45	12	99
Special Districts, School Boards, and other[c]	22	64	12	98
Total	**67**	**109**	**24**	**200**

Source: Calculations and categorization by CRS. Raw data provided by FEMA staff, and were current as of June 15, 2012.

Notes: This table does not count multiple loans that were issued to the same government entity for one disaster, as allowable under the 2008 and SCDL rules.

a. This category includes U.S. territories.

b. This category includes general purpose municipal city, county or parish governments.

c. This category includes government entities such as water and sewage boards, transportation special districts like maritime ports and airports, hospitals, sheriff's departments, and school districts/boards.

TCDLs can be approved for a local government in either the fiscal year in which the disaster occurred or the immediately following fiscal year. Only one TCDL may be approved for any one local government as the result of a single disaster.[25] Also, as a result of the Disaster Mitigation Act of 2000 (P.L. 106-390, henceforth DMA 2000), a loan cannot be issued to any local government that is in arrears in payment on a previously existing community disaster loan.[26]

The regulatory criteria for what constitutes a substantial loss in tax or other revenues, and how a local government demonstrates a need for the financial assistance, are considerably more complex. There are two factors used by FEMA to decide if there is a substantial loss in tax or other revenues.[27] The first is whether the reduction in revenue is severe enough to significantly and adversely impact the level of essential municipal services being provided by the local government. The second is whether the disaster has caused a projected 5% loss in total revenue for the fiscal year of the disaster or the fiscal year following the disaster. This particular eligibility criterion has often been altered in statute to increase the percentage of revenue loss required to receive loans of larger sizes, and is one of the key differentiating elements of the special and 2008 loans.

FEMA regulations contain eight different factors that are used to determine if a local government has a need for financial assistance.[28] The factors include whether the local government is in danger of municipal insolvency, if there are available cash or liquid assets to cover the losses, and the existing debt ratio of the local government. These regulatory factors are consistent across the traditional, special, and 2008 loans. Some of the factors are designed to gauge the current financial health of the government, and thereby the need for financial assistance. These provisions, if implemented strictly, may inhibit the loan eligibility of fiscally sound jurisdictions and enhance the loan eligibility of fiscally weak jurisdictions.[29]

Loan Application

To help implement these eligibility criteria, FEMA has established a formal application process for the program.[30] The regulations specify that the state in which the local government is located review and validate the application, with the Governor's Authorized Representative (GAR) officially approving the application.[31] One of the application forms requires the local government to self-certify that it meets a number of eligibility terms, to include most of the factors used to determine a financial need for the loan.[32]

The local government is also required to provide its most current financial reports for the three fiscal years prior to the disaster, in addition to other documentation required by FEMA, to support its application.[33]

To help ensure that a loan is repaid, the state must co-sign the promissory note for the loan if a loan application is accepted. If the state cannot legally sign the note, the local government must pledge collateral security to cover the principal amount of the note.[34] These requirements were newly introduced to the program in the 1988 rulemaking. In that rulemaking, FEMA claimed these changes were necessary to counteract local governments that did not always act in a "fiscally responsible manner, and have been reluctant, and in some cases refused, repeatedly, to repay the uncancelled principal and related accrued interest after the regulatory loan cancellation and appeal process had been exhausted."[35] FEMA staff have indicated that, since this addition to the regulations, there have been unspecified circumstances in which states have been unwilling to co-sign a loan application with a local community. In these circumstances, local governments have either provided collateral security or withdrawn their loan applications.[36]

During the application process, FEMA reviews whether or not a government experienced a five percentile loss in total tax and other revenues in either the fiscal year of the disaster or the following fiscal year. As mentioned previously, this eligibility factor is part of the substantial loss determination required in statute. Often with the assistance of contract staff hired by FEMA, the local government is required to extensively document empirical evidence of a 5% predicted revenue loss in lieu of actual revenue loss.[37] For the purposes of evaluating a loan application and the eligibility of a local government, FEMA will not consider any voluntary reduction in the collection or assessment of tax or other revenues as a "legitimate" revenue loss to meet the 5% standard.[38] This prevents a government from willfully reducing its revenues in order to become eligible for a loan, or to increase the overall loan size once eligible. The applicant is also allowed to submit a formal letter to FEMA explaining the financial context of its request, and its pressing need for the financial assistance. For a demonstration of need, FEMA presumes the need exists through the self-certification of the government, unless evidence is found that would otherwise contradict eligibility on this provision.

FEMA reviews and notifies each applicant whether its loan has been accepted. If denied a loan because inadequate information was provided, the local government is allowed to resubmit a revised application within 60 days.[39]

FEMA staff said that in practice they informally pre-screen communities for their eligibility for CDLs when working with them in the disaster recovery process following a major disaster. Therefore, FEMA could not provide any instances when an applicant that formally submitted an application had been denied. The informal pre-screening process makes it difficult to ascertain what are the most common factors that prevent a government from applying for or receiving a loan. However, based on anecdotal evidence, FEMA staff suggested that local governments often do not have the 5% in revenue loss to justify a substantial loss from the disaster. It is also unclear whether local governments with strong financial standing are any more or less likely to receive a loan, as hypothetically possible under the need determination criteria.[40]

Loan Size

From FY1974 to FY2000, the only statutory limitation on the size of a traditional community loan was that it could not exceed 25% of the annual operating budget of the local government for fiscal year in which the major disaster occurred.[41] As a result of an amendment made in the DMA 2000 (P.L. 106-390), the dollar amount of any loan was further limited to $5 million per loan, regardless of what percentage of the annual operating budget that figure represented.[42] One of the stated intentions of the DMA 2000 was to reduce the financial cost of disaster assistance following natural disasters.[43] The $5 million cap was intended to prevent some of the extremely large loans, such as one for $127.2 million to the Virgin Islands after Hurricane Marilyn in 1995, from being issued and ultimately cancelled.

The $5 million cap has been a source of controversy since inception. Many argue that it inequitably limits the amount of assistance to local governments with large operating budgets versus those with smaller budgets. Following the 9/11 terrorist attacks, a bill was introduced in the House to remove the cap and make other revisions to the CDL program, but it failed to reach the floor.[44] The $5 million provision was one of the elements Congress waived, under certain conditions, by special legislation for the SCDLs and 2008 CDLs. The cap remains in place for all future disaster loans unless otherwise stipulated by Congress. No loans were issued between the passage of the $5 million cap and its subsequent waiver for the SCDLs (discussed further later), though the $5 million cap did restrict the size of 2008 CDLs under certain conditions and was in place for four loans issued under traditional regulations for disasters in 2007.[45]

Through regulation, FEMA further limits the size of a loan to a local government's projected need for financial assistance. Using the financial documents provided in the application, FEMA calculates the cumulative amount of projected revenue loss plus projected unreimbursed disaster-related expenses (UDREs) for the fiscal year of the disaster and the following three fiscal years.[46] The projected sum over the three-year period serves as the size of a loan if it is less than the 25% of the annual operating budget or $5 million dollar cap. The goal behind the regulation is to prevent a community from taking out a loan in excess of its projected need.

The SAFE Port Act (P.L. 109-347), enacted on October 13, 2006, further raised the loan size limit for a traditional CDL from 25% to 50% of a local government's annual operating budget in the fiscal year of a disaster. This increased loan size can only be issued if a government's tax and other revenue loss is equal to or more than 75% of its annual operating budget. No government, including those impacted by Hurricanes Katrina and Rita, has received a loan using this standard. This is because it is rare to lose revenues up to 75% of an operating budget following a disaster. Moreover, were this stipulation applied to a future loan, the size would still be limited by FEMA's calculation for projected need, and therefore might not reach a 50% operating budget figure.

Interest Rates

The CDL statute is silent on what interest rate should be charged for loans issued under the program. As a default interpretation, FEMA has set interest rates for all TCDLs, regardless of size, to the rate for five-year maturities as determined by the Secretary of Treasury on the date of execution.[47] The rate is fixed for the course of a loan. The interest rate on CDLs is often higher than the average rate on municipal (state and local) bonds of similar maturity. This is because the federal tax exemption of interest on state and local government bonds enables those governments to sell bonds at lower interest rates than comparable federal bonds. Theoretically, the relatively higher CDL rate implies that localities with strong credit ratings may be better off borrowing from the private credit market, if they were permitted to borrow to cover operating expenses.[48] Communities with a weaker credit rating—or those anticipating a sufficient revenue loss to justify loan cancellation—may be more attracted to traditional CDLs. The average interest rate for all TCDLs since program inception is 6.59%.

Administration

Though the total amount of the loan is obligated by FEMA, the local government must request disbursement of the loan in increments based on a defined schedule or by need. With each increment requested, the local government is required to submit new financial evidence, which FEMA considers, to justify the withdrawal.[49] In implementation of the regulation, FEMA has stated that it rarely denies disbursement requests from local governments because the requests are typically justified by financial evidence.[50]

Use of Funds

The only guidance found in the statute on the use of the loan funds is that a local government must use the funds "to perform its governmental functions."[51] FEMA has interpreted this clause to further mean that the funds can only be used to carry on existing local government functions of a municipal operation character or to expand such functions to meet disaster-related needs. In an interim rule released for the SCDLs, FEMA indicated further that the local government may pass loan proceeds to a private non-profit that provides government services, though the responsibility for repayment remains with the local government.[52] The funds cannot be used to finance capital improvements or for the repair or restoration of damaged public facilities. Also, the funds cannot be used to fund the "match" share for another federal assistance program.[53] FEMA monitors the use of the funds in the annual reports submitted by the local governments and through the requests for disbursements.

Summary of Traditional Loans

FEMA has approved 68 TCDLs to 68 different local governments from 13 different states and 2 different territories. Of those loans, only 57 loans were used by the local governments, meaning they borrowed some dollar amount from available funds. In total, local governments were approved for $284.1 million, which is $48.1 million more than the total $236 million that has been borrowed. **Table 6** summarizes the amount of loans borrowed through the traditional regulations by government type.

Cancellation of Traditional Loans

By statute, repayment on a loan may be cancelled in whole or in part if the "revenues of the local government during the three full fiscal-year period following the major disaster are insufficient to meet the operating budget of the local government, including additional disaster-related expenses of a municipal operation character."[54] Though it has been amended, this provision's intention was included in the original authorization in 1974.[55] FEMA will also forgive any amount of related interest owed on the cancelled principal of a loan.[56]

To implement this statute, FEMA uses the regulations in 44 C.F.R. §206.366 to determine how much, if any, of a loan should be cancelled. Typically, FEMA hires an outside auditing firm to perform the required analysis for the forgiveness evaluation of a community's operating budget. FEMA's cancellation decisions are based on the auditing firms' analysis of a local government, as well as any additional explanatory information that the jurisdiction provides in support of its cancellation application.

Table 6. Borrowed TCDL Amounts by Government Type

Government Type	Number of Loans	Total $ Amount of Borrowed Principal, in Millions	Average $ Amount of Borrowed Principal per Loan, in Millions
Territory[a]	3	$187.48	$62.49
Municipal Governments[b]	37	$46.16	$1.25
Special Districts, School Boards, and other[c]	17	$2.36	$0.14
TOTAL	**57**	**$236.00**	**$4.14**

Source: Calculations and categorization by CRS. Raw data provided by FEMA staff, and were current as of June 15, 2012.

Notes: This table does not include loans where zero dollars were borrowed or the loan was withdrawn by the choice of the local government (11 loans excluded). Dollar figures have been rounded to nearest hundredth of a million. Totals may not add due to rounding error.

a. This category includes loans made to U.S. territories, namely American Samoa (1) and the Virgin Islands (2).

b. This category includes general purpose municipal city, county, or parish governments.

c. This category includes government entities such as water and sewage boards, transportation special districts like maritime ports and airports, hospitals, sheriff's departments, and school districts/boards.

Review Process for Loan Cancellation

To be eligible for loan cancellation, a local government must first submit a cancellation application before the expiration date of the loan through its state GAR and the applicable Regional Administrator for FEMA.[57] This means a local government must apply for cancellation within five years from the date the promissory note was signed.[58] As a matter of practice, FEMA notifies local governments of the cancellation application's requirements, and begins assisting them early in the process to ensure timely delivery.

An extensive amount of supporting financial documents is required by FEMA to review a cancellation application. These documents are listed in full in 44 C.F.R. 206.366(c)(1), and include items that help FEMA understand the local government's property tax evaluation practices and individual fund balance sheets. In addition to the documents, FEMA encourages applicants to submit a written narrative explaining their financial situation to make it easier for FEMA to understand the unique circumstances of each local government.[59]

Upon receipt of an application, FEMA begins its evaluation process. Succinctly, the following steps are followed in the evaluation:

1. FEMA determines if a cumulative operating deficit exists in the three fiscal years after the disaster. Without a cumulative operating deficit, no loan cancellation is possible. A "cumulative operating deficit" is essentially the shortfall between actual revenues and actual expenditures of a local government over the three-year period.

2. FEMA then tries to associate the deficit to either a disaster-related loss in revenue or an increase in expenditures due to unreimbursed disaster-related expenses. In this step, FEMA is essentially trying to determine what amount of the deficit has been "caused" by the disaster, and not by other factors.

3. After step 2 is complete, a total amount of disaster-related deficit will have been calculated. FEMA may cancel an equivalent amount in loan principal owed by the local government (either partially or in full). The associated interest of the cancelled principal is also forgiven.

This process is further explained in the rulemakings for the SCDLs, as well as a supporting presentation made by FEMA explaining cancellation for SCDLs.[60] The process outlined should not be distinctly different for traditional loans, as their guiding regulations are the same except for the additional clarifications explained in "Cancellation of Special Community Disaster Loans." This process is difficult to implement in practice because of the numerous ways in which operating budgets, revenues, and additional disaster-related expenses of a municipal operating character can be interpreted. The relevant regulations on how FEMA interprets these terms for TCDLs are discussed immediately below.

Any applicants that are denied cancellation of all or part of their loan may appeal the decision. In practice, this appeal is handled within the same division/directorate in FEMA, but by more senior staff. An appeal must be submitted within 60 days of a denial. All appeal decisions are final.[61]

Operating Budget

"Operating budget" is defined as "actual revenues and expenditures of the local government as published in the official financial statements of the local government."[62] The term "actual" is important because not all budgeted monies are always spent, and not all projected revenues are collected, in any fiscal year. FEMA is only interested in what actually occurred during the full three-year period, as opposed to annual projections. Also, since local governments often have budget deficits prior to the disaster, FEMA does not consider it within the purpose of the CDL program to fund these deficits. Therefore, when calculating the operating budget of a local government, FEMA reduces the budget size by the pre-disaster deficit amount.[63] Further, since the funds are meant to be used only for operating expenses and not for capital expenses, any amount of money the government transfers from an operating budget to a capital budget is added back into the total operating budget.

Revenues

The term "revenues" is not clearly defined in the traditional regulations. However, FEMA does provide key stipulations on how they should be calculated. If a local government reduces the tax or other revenue rates for undamaged property, such as reducing the percentage charged for property tax or the fees charged for utilities, FEMA uses the rates from the pre-disaster level in its overall calculation for cancellation.[64] Many governments will reduce taxes and fees on unaffected areas or experience a decrease in property values in unaffected, but nearby, areas. Through this regulation, FEMA uses the pre-existing rates for revenue calculation and appraisals of property value, as if the rates/appraisals were never reduced. This regulation prevents hypothetical situations where a government could willfully reduce its revenues, thus encouraging larger loan sizes and cancellation rates. As noted in the traditional rules, this regulation "may result in decreasing the potential for loan cancellations."[65] The regulations do, however, leave open the possibility that taxes and fees could be reduced on damaged property, and that FEMA could credit a loss in revenue associated with lower appraisals for damaged property.

Disaster-Related Expenses of a Municipal Operating Character

As with the projections of revenue loss during loan size calculations, FEMA will include "unreimbursed disaster-related expenses" (UDREs) of the local government in the overall amount that the loan may be cancelled. UDREs include

> those [expenses] incurred for general government purposes, such as police and fire protection, trash collection, collection of revenues, maintenance of public facilities, flood and other hazard insurance, and other expenses normally budgeted for the general fund.... [66]

Expenses that are not eligible for forgiveness include expenditures associated with debt service; any major repairs, rebuilding, replacement, or reconstruction of public facilities or other capital projects; intragovernmental services; special assessments; and trust and agency fund operations. Any expenses that may be eligible for coverage under another federal assistance program are also ineligible, as a means of preventing double dipping for federal assistance.[67]

Loan Repayment

The normal term of a CDL is five years, where the full principal and accumulated interest are due all together at the end of the five-year term on whatever amount is not cancelled. FEMA may consider requests for an extension on the repayment of the loan, based on the local government's financial condition. However, the total term of a loan normally may not exceed 10 years, except under extenuating circumstances outlined in the regulations.[68] In these extenuating circumstances, an additional loan may be issued by FEMA after the first 10 years, allowing a second 10-year promissory note to be signed. In order to receive this additional amount of time to repay a debt, a local government must first attempt to apply for credit in the private market at a rate equivalent to the current Treasury rate for a similar loan. If additional time is granted, the outstanding principal and interest becomes the principal amount of a new loan from FEMA. There is currently one loan outstanding past the 10-year repayment mark under these circumstances. Local governments may also make prepayments on a loan without penalty.[69] In the event of default, FEMA may request administrative offset against other federal funds due to the borrower or refer the loan to the Department of Justice for enforcement and collection.[70]

Summary of Traditional Loan Cancellation Rates

The total dollar amounts and percentage rates of cancellation for TCDLs are displayed by government type in **Table 8** and **Table 9**. To summarize, of loans that have had any amount of money borrowed, 46.4% of the loans were at least partially forgiven. On average, each loan had 38.9% of its principal cancelled by FEMA. The total percentage of principal cancelled was 97.2%, largely due to the weighted impact of several large loans having been cancelled. In addition to the loans that have been cancelled through normal regulations, one large loan made to the U.S. Virgin Islands following Hurricane Hugo in 1989 was cancelled in part by special legislation. Through the normal cancellation regulations, FEMA forgave $21 million of the $50.1 million in principal borrowed through the loan (and in doing so, also forgave the associated interest on the $21 million). The U.S. Virgin Islands also repaid about $7.7 million in interest on the remaining principal. The remaining balance of principal and interest was cancelled through legislation. In the Department of the Interior and Related Agencies Appropriation Act, 2001 (P.L. 106-291), Congress transferred funds from the Department of the Interior to FEMA for the purpose of cancelling some of the remaining interest on the U.S. Virgin Island loan.[71] The following year, in the Department of the Interior and Related Agencies Appropriation Act, 2002 (P.L. 107-63), Congress again transferred funds from Department of the Interior to FEMA, this time for the purpose of cancelling the remaining balance on the loan.[72] By doing so, Congress cancelled the U.S. Virgin Islands' obligation to pay a remaining $27 million in principal and $17.5 million in interest.

SPECIAL COMMUNITY DISASTER LOANS FOR HURRICANES KATRINA AND THE 2005 HURRICANE SEASON

In addition to the heavy loss of lives and the dislocation of hundreds of thousands of families, Hurricanes Katrina and Rita caused devastating damage to property and seriously disrupted the economic activity that normally provided the tax and revenue base of the affected areas, especially in Louisiana and Mississippi. As the full scale of the devastation was revealed, the 109th Congress passed a series of emergency supplemental appropriations to provide financial assistance to the region. Within two weeks, Congress had passed two separate emergency supplemental appropriations to meet the needs arising from Hurricane Katrina (P.L. 109-61 and P.L. 109-62). These appropriations focused on funding more commonly used disaster assistance programs and accounts, such as the Disaster Relief Fund (DRF), and did not initially include any appropriation for the Disaster Assistance Direct Loan Program (DALDP) account for the CDL program. However, on October 7, 2005, the 109th Congress passed and President Bush signed into law the Community Disaster Loan Act of 2005 (P.L. 109-88, henceforth CDL Act of 2005). The act transferred up to $750 million to the DALDP account from the $50 billion previously appropriated for disaster assistance following Hurricane Katrina in P.L. 109-62. The $750 million was available to support up to $1 billion in special loans to local governments until expended. The CDL Act of 2005 also allowed for an additional $1 million of the disaster relief funds provided by P.L. 109-62 to be transferred to the DALDP account for the administrative expenses of the program. Loans totaling the full $1 billion were approved by FEMA under the CDL Act of 2005. Since the first obligation was quickly exhausted, the 109th Congress passed the Emergency Supplemental Appropriations Act for Defense, the Global War on Terror, and Hurricane Recovery, 2006 (P.L. 109-234, henceforth Emergency Supplemental of 2006) on June 15, 2006. The Emergency Supplemental of 2006 transferred another $278.8 million to the DADLP to support an additional $371.733 million in direct loans to communities affected by Hurricane Katrina or other hurricanes in the 2005 season (such as Hurricane Rita). An additional $1 million was appropriated for administrative expenses. The numerical relationship between the appropriated amounts in each Act ($750 million and $278.8 million) and the allowable amount of loans ($1 billion and $371.733 million) was based on the assumption of a 75% credit subsidy rate for the loans.[73]

From the possible $371.733 million available under the Emergency Supplemental of 2006, FEMA approved loans totaling $271 million for all eligible applicants, leaving $101 million of the loan authorization unused. Though collectively the loans from both laws are called special community disaster loans (SCDLs), each law had slightly different statutory provisions for how the loans could be issued and administered by FEMA.

Distinguishing Features from Traditional Loans

Purpose of the Special Loans

The intended purpose of the SCDLs was slightly different than the purpose of the TCDLs. The CDL Act of 2005 appropriated the initial funds for the loans "to assist local governments in providing essential services."[74] The Emergency Supplemental of 2006 echoed this purpose statement. The impact of this clause is reflected by a difference in language in the opening sections of the regulations for each set of loans. In regulations for the TCDLs, it states that the program's purpose is for "... any local government which has suffered a substantial loss of tax or other revenues as a result of a major disaster and which demonstrates a need federal financial assistance *in order to perform its governmental functions*" (italics added).[75] This language directly replicates the traditional purpose statement found in the opening portions of Section 417(a) of the Stafford Act. In contrast, the regulations that applied to SCDLs state "... in order to provide essential services." In an initial interim rule issued shortly after the CDL Act of 2005, FEMA noted that by using the "essential services" definition it expected "proceeds from these loans will be limited to the performance of core municipal operating functions."[76] Further, in the Notice of Public Rulemaking (NPRM) for the special regulations issued in 2009, FEMA cited this difference between "essential services" versus "governmental functions" as being restrictive to the purpose of the special loans by comparison to the TCDLs.[77] FEMA also indicated that local government entities such as recreational districts were unable to apply for and receive loans under the essential services definition.[78] Outside of this restriction, it is unknown what other practical impacts, if any at all, this difference had on eligibility or the cancellation of SCDLs.

Eligibility

The SCDLs from the CDL Act of 2005 were available to any local government for the purpose of providing essential services.[79] Though many local governments under various disaster declarations could have applied for loans, the full $1 billion went to governments in Louisiana or Mississippi under major disaster declarations for Hurricane Katrina.[80]

Eligibility for the SCDLs made available in the Emergency Supplemental of 2006 was more restrictive, and was limited to governments under major disaster declarations from Hurricane Katrina and the 2005 hurricane season.[81] Under this limitation, only governments in Louisiana and Mississippi applied for and received loans through this appropriation. Local governments in Alabama, Florida, and Texas could have also applied for and received these special loans, but did not. Because eligibility for these loans was limited to the 2005 hurricane season, FEMA interpreted that the Emergency Supplemental of 2006 loans must be made by September 30, 2006, the end of FY2006, in accordance with the existing regulatory provision that loans can only be issued in the fiscal year of the disaster or the succeeding fiscal year (FY2005 or FY2006).[82] Since the declarations for Hurricane Rita were issued at the very end of FY2005 (on September 24, 2005),[83] there was significant time pressure on the application process for the loans issued under Emergency Supplemental of 2006.

Number of Loans Issued to a Local Government

As with traditional loans, the issuance of SCDLs was restricted to the fiscal year of the disaster declaration or the succeeding fiscal year (FY2005 or FY2006). However, the stipulation that a government may only receive one loan per disaster was removed to account for two loan appropriations made by the 109th Congress that had separate eligibility provisions.[84] FEMA anticipated that the first appropriation under the CDL Act of 2005 would be insufficient to meet the full eligible demand from local communities impacted by the hurricanes. Therefore, loan amounts were allocated to as many applicants as equitably as possible until funds were exhausted, in part by prioritizing some local governments that provided more essential services (such as hospitals and school boards) over governments providing fewer essential services.[85] In some cases, this did not allow all local governments to receive their full loan eligibility amount, so, upon the second appropriation (via the Emergency Supplemental of 2006), additional loans were issued to some communities to meet their full eligible amount under the CDL Act of 2005. An additional loan could have also been issued to local governments under the unique loan size provisions of the Emergency Supplemental of 2006 or for damage from Hurricane Rita as well as Hurricane Katrina.

Loan Size

As applicable to traditional loans issued prior to 2000, the CDL Act of 2005 removed the restriction on the $5 million cap for loans, regardless of revenue loss. However, the loans issued under the CDL Act of 2005 were still restricted to 25% of the annual operating budget of a government.

The $5 million cap was removed again in the Emergency Supplemental of 2006. In addition, loans were allowed to reach 50% of the operating budget for a local government in the fiscal year of the disaster. However, this uniquely large loan size was only allowed if the local government had suffered a projected loss of 25% or more in tax revenues. Because the percent-of-budget limit was raised to 50%, communities that applied in the first round for SCDLs under the CDL Act of 2005 for up to 25% of their operating budget could now apply for an additional 25% under the emergency supplemental loans. FEMA still maintained the regulatory provision that loan size was limited to the projected revenue loss plus unreimbursed disaster expenses,[86] so it is unclear how many loans actually reached or approached the full 50% standard. Under the CDL Act of 2005, FEMA issued a total of 136 loans (84 to Louisiana and 52 to Mississippi). Under the Emergency Supplemental of 2006, FEMA issued 16 additional loans (12 to Louisiana and 4 to Mississippi).[87] However, in the information provided to CRS by FEMA, loans were not specifically identified by appropriation, so it is unclear which loans were issued under which law. It is also unclear if additional loans issued under the Emergency Supplemental of 2006 were issued to new or old recipients, or what the size of the loan was as a percent of the local government's operating budget.

"Tax and Other" Revenue Clause

Section 417(a) of the Stafford Act stipulates that loans are available to governments that have suffered "a substantial loss of tax and other revenues." As with the TCDLs, special loans issued under the CDL Act of 2005 followed this provision. However, under the Emergency Supplemental of 2006, a local government could receive a loan for up to 50% of their operating budget if they suffered a loss of 25% or more in tax revenues due to Hurricane Katrina or Hurricane Rita.[88] This distinction is important because many local government entities, especially special districts, rely on revenues other than taxes (such as charges and fees) and therefore had difficulty meeting the standard of a 25% loss strictly from taxes. These local government entities included, for example, hospitals, ports, airports, regional transit agencies, and communications authorities.

Interest Rates

As with traditional loans, the statute for the SCDLs included no specific provisions regarding the interest rate for the loans. By default, FEMA created regulations setting the interest rates on the SCDLs to the rate for five-year maturities as determined by the U.S. Treasury. However, possibly as a consolation offered to those concerned about the non-cancellation provision for the SCDLs,[89] FEMA also allowed interest rates to be reduced "if an applicant can demonstrate unusual circumstances involving financial hardship."[90] The reduced interest rate was the U.S. Treasury's five-year maturity rate plus one percentum, adjusted to the nearest 1/8%, and reduced by one-half. For example, assume that the yield on five-year Treasury bonds were 4.32%, as it was on October 21, 2005. Adding one percentum would give 5.32%. Rounding that to the nearest 1/8% would give 5 3/8%. Reducing that by one-half would give 2 11/16% (2.69%) as the subsidized interest rate on SCDLs. It is unknown how applicants demonstrated to FEMA they had "unusual circumstances involving financial hardship," but all SCDLs were released with this lower interest rate. Some might argue that the extreme devastation of Hurricane Katrina was self-evident justification for applying the lower interest rate. The average interest rate for the 157 approved SCDLs was 2.86%.[91]

Summary of SCDL Loans

Table 7 summarizes the amount of loans borrowed through either the CDL Act of 2005 or the Emergency Supplemental of 2006. In total, 157 separate SCDLs were approved by FEMA for 109 separate local governments in Louisiana and Mississippi. However, local governments either withdrew or did not elect to borrow money in 28 of the loans. Therefore, in terms of only those loans that had some amount of borrowed principal, 129 loans were issued and used by 92 local governments. In total, local governments were approved for $1,270 million in principal, which is $230 million more than the $1,040 million that was borrowed. FEMA issued 53 special disaster loans that exceeded the previous limit of $5 million, with the largest being two different loans for $120 million each to the City of New Orleans. On average, the local governments in Louisiana borrowed more than those in Mississippi, especially in the municipal government category. Presumably, this difference was due to the scope of the damage in Louisiana and the size of the local government budgets.

Cancellation of Special Community Disaster Loans

The CDL Act of 2005 contained a unique provision preventing the loans issued from being cancelled under Section 417(c)(1) of the Stafford Act.[92] Though the stipulation against loan forgiveness was controversial, it was reportedly insisted upon by the Office of Management and Budget and the Republican leadership in the House as a condition for providing the loan assistance.[93] Several Members made statements on the House and Senate floors objecting to the requirement that the loans be repaid in full, without the possibility of cancellation.[94] Thirteen days after the enactment of the CDL Act of 2005, companion bills were introduced in each chamber of Congress to repeal the provision that disallowed the cancellation of the special CDLs, but neither was voted upon in the 109th Congress.[95] The loans authorized by the Emergency Supplemental of 2006 also prohibited loan cancellation. Attempts to repeal the restriction on loan forgiveness were taken up again in the 110th Congress. Some of the bills only attempted to remove the provision for the loans in the CDL Act of 2005,[96] while others also attempted to allow forgiveness for the loans from the Emergency Supplemental of 2006.[97] Ultimately, the U.S. Troop Readiness, Veterans' Care, Katrina Recovery, and Iraq Accountability Appropriations Act, 2007 (P.L. 110-28) allowed cancellation for both the loans under the CDL Act of 2005 and the Emergency Supplemental of 2006.[98] In response, FEMA initiated a rulemaking process to allow for the SCDLs to be forgiven.

Background on Special Community Disaster Loan Regulations

Table 3 displays the key steps in the rulemaking process associated with the SCDLs. The interim rule, released in October 2005, was issued under the presumption that SCDLs would not be eligible for cancellation (per the original statute of the CDL Act of 2005). Therefore, the interim rule essentially replicated existing regulations for the CDL program, except that it removed the $5 million cap on loan size; restricted loan use to "essential services"; and removed provisions on how loans could be cancelled.[99]

Table 7. Borrowed SCDL Amounts by State and Government Type

State	Government Type	Number of Loans	Total $ Amount of Borrowed Principal, in Millions	Average $ Amount of Borrowed Principal per Loan, in Millions
Louisiana	**Municipal Governments**[a]	22	$349.0	$15.9
	Special Districts, School Boards, and other[b]	57	$488.3	$8.6
Mississippi	**Municipal Governments**	21	$95.3	$3.3
	Special Districts, School Boards, and other	29	$108.1	$5.1
	TOTAL	**129**	**$1,040.7**	**$8.1**

Source: Calculations and categorization by CRS. Raw data provided by FEMA staff, and were current as of June 15, 2012.

Notes: This table does not include loans where zero dollars were borrowed or the loan was withdrawn by the choice of the local government (21 loans excluded for LA, 7 excluded for MS). Dollar figures have been rounded to nearest tenth of a million. Totals may not add due to rounding error.

a. This category includes general purpose municipal city, county, or parish governments.

b. This category includes government entities such as water and sewage boards, transportation special districts like maritime ports and airports, hospitals, sheriff's departments, and school districts/boards.

After the change in law in May 2007 that allowed SCDLs to be cancelled (P.L. 110-28), FEMA proposed a rule in April 2009 indicating how SCDLs would be processed for possible cancellation. According to media reports, the time delay between the passage of the law allowing cancellation and the release of the proposed rulemaking (approximately 22 months) was criticized by many lawmakers.[100] The April 2009 release date for the proposed rule was approximately 18 months ahead of the five-year maturity date for the first batch of SCDLs, though it came about six months after the first loans could have technically been reviewed for cancellation (after the first three fiscal year period since issuance). In essence, the proposed rule amended the existing SCDL rules by adding the identical cancellation regulations used for traditional community disaster loans since 1988.[101] According to FEMA, the agency contemplated "possible changes to these well-established cancellation procedures" but found that

> the cancellation provisions for the traditional Community Disaster Loan program work— they provide sufficient and accurate information on which FEMA can base its decision to cancel loans—and compliance on the part of the borrower is relatively easy. There are no significant issues with these existing procedures that require revision....[102]

The final rulemaking allowing for SCDL cancellation was released in January 2010, which was approximately nine months before the maturity date of the first SCDLs. FEMA received 68 comments on the proposed rule from 2009, and in response to these suggestions made five substantive changes to the regulations.[103] In addition to comments from the general public and representatives of local governments who received special loans, several past and present Members of Congress submitted comments on behalf of their constituents.[104] The regulations for SCDL were changed in the final rule by:[105]

- Altering how property tax revenue loss is evaluated for cancellation applications. In principle, this amendment made cancellation of loans more likely in certain restricted circumstances. Additionally, FEMA clarified how it would evaluate the loss of property revenue in general. In implementing this regulation, FEMA indicated that

  > unless provided information to the contrary, FEMA will assume that any assessed property value decline during the three full fiscal years after the disaster was related to the disaster, and not to general market

conditions, as market conditions themselves were severely affected by the disaster during that period of time.[106]

- Clarifying the definitions of "revenues" and "operating expenses."[107] The regulation continued to rely on the existing definition of operating budget. These definitions were developed from the Government Accounting Standards Board and published by the Government Finance Officers Association.
- Identifying the different persons/positions within FEMA that would rule on the initial cancellation application and any possible appeal. Although these positions were different, they were within the same management chain at FEMA (with the appeal being ruled on at a higher level).
- Providing a specific timeline for the cancellation review process, allowing FEMA 60 days to process the evaluation.
- Allowing local governments to submit financial data for the three full fiscal years following the disaster, as normal, or the 36 calendar months following a disaster. This aided local governments whose fiscal years did not match the typical October 1st start date of the federal fiscal year.

These revisions aside, FEMA reiterated that the general procedures for loan cancellation used for TCDLs would be repeated, having

> found them to be an efficient and accurate method of determining when revenues of a local government are insufficient to meet its operating budget. These procedures were successfully applied after other major hurricanes, including but not limited to hurricanes Andrew (1992) and Marilyn (1995).[108]

Single Cancellation Review per Government Locality

As discussed above, local governments often received more than one SCDL loan from FEMA. Though some of these loans were distinct in the sense that they were issued under different appropriations and allowances under the law (CDL Act of 2005 and the Emergency Supplement of 2006), FEMA completed a single cancellation review process per local government, regardless of how many special loans a government received. The local government information needed for FEMA to process each loan cancellation application was the same if the government had four loans or one. In theory, the cancellation review process was conducted in the same manner as for the TCDLs,[109] except that

- Clarified definitions were applied in the review of local government budgets;
- The loss of property taxes was handled more flexibly to the benefit of local governments; and
- A definitive timeline for the application review and appeal process was followed by FEMA.

FEMA contracted with a private accounting firm to help it conduct the application review. In addition to the regulations, FEMA also provided local governments with several additional support documents to help guide the governments through the cancellation application process, including a review of FEMA's accounting methodology. After calculating a total dollar amount of operating budget deficit that is disaster-related, FEMA cancelled an equal amount of money from the total principal owed by each local government, cancelling the principal balances loan by loan. For example, consider a hypothetical local government that had $20 million in "forgivable" operating budget deficit. If the local government had two loans, one for $15 million and another for $10 million, FEMA would cancel the full principal and interest of the first loan and $5 million of the principal and associated interest from the second loan.

Controversy Surrounding SCDL Cancellation

Historically, decisions rendered by FEMA for cancelling community disaster loans have provoked controversy, especially for particularly large loans.[110] Decisions rendered by FEMA for cancelling SCDLs were no exception. The majority of the local governments that did not receive full cancellation proceeded to appeal FEMA's initial decision, and 10 local governments were continuing their appeals as of June 15, 2012. The specific financial audit procedures used by FEMA to calculate the cancellable amount of a loan have been particularly contentious. FEMA's intention was to apply standard practices and accounting methods for reviewing the necessary elements (operating budget, revenues, expenditures, etc.) established by the Government Accounting Standards Board in hopes of avoiding confusion or inconsistencies. However, Senator Mary Landrieu wrote to the U.S. Attorney General, Eric Holder, enumerating a number of issues with FEMA's cancellation procedures.[111] The Louisiana legislative auditor also conducted a review of FEMA's procedures, and found several ambiguities in how FEMA would treat certain types of revenues, expenditures, and budgetary procedures.[112] Some of these ambiguities were claimed to have led to inconsistencies in the review of the cancellation application. CRS cannot independently verify or evaluate these assertions regarding the financial audit procedures of FEMA. Others, including Senator David Vitter, further suggested that FEMA's cancellation procedures unfairly treated the governments that "slashed budgets" because they do not ultimately have a three-year operating budget deficit, even though their communities were still recovering and experiencing significant revenue shortages.[113]

Comments made by Vice President Joseph Biden in a public speech in St. Bernard's Parish, LA, on January 15, 2010, subsequently increased attention around the cancellation of SCDLs. Local media sources quoted Mr. Biden as saying to the crowd, in reference to the SCDLs, "I advise you to apply quickly, because when you apply you are going to get the right answer ... . You're going to get your money."[114] Other sources further quoted Mr. Biden as saying "It sure looks like to me you're still rebuilding, so I want to tell you something, it will be ... [verbal pause].... You're gonna get the money."[115] The Vice President's comments were perceived by some community members, state legislators, and Members of Congress as a promise or guarantee that the SCDLs would be cancelled in large or completely. The Vice President's comments on that occasion have since been used to argue for greater flexibility on behalf of FEMA to more readily cancel the loans.[116]

Summary of SCDL Cancellation Rates

The appeal and adjudication process for special loan cancellation continues, with a total of 17 loans issued to 10 different local governments still in appeal as of June 15, 2012. For the purposes of the calculations presented in **Table 8** and **Table 9**, loans still under appeal were categorized by their financial status as of June 15, 2012, as provided by FEMA. If, for instance, the appealed loans have already had some amount cancelled, but the local governments are appealing for more forgiveness, the loan is categorized as partially cancelled. Of the loans in appeal, four have already been partially cancelled. A particular challenge in assessing SCDL cancellation rates is that multiple loans were often issued to a single local government entity, though FEMA ultimately conducted one cancellation review per local government, as opposed to per loan issued (see "Single Cancellation Review Per Government Locality"). For the purposes of evaluating overall cancellation rates, some of these loans should ideally be combined, while others should be treated separately, depending on why each loan was issued. For instance, if a second loan was issued for the sole purpose of providing the local government its full eligible amount under the CDL Act of 2005, these loans should probably be treated as a single loan. But if an additional loan was issued to a local government for unique damage caused by Hurricane Rita and not Hurricane Katrina, or for loan size provisions under the Emergency Supplemental of 2006 and not the CDL Act of 2005, it might be proper to statistically treat them as separate loans. However, too little information is available to CRS to conduct analysis of these loans along these suggested categories. As a default, these loans are treated as separate loans per their issuance by FEMA.

To summarize from **Table 8** and **Table 9**, 59.7% of the SCDLs were at least partially forgiven. On average, each special loan had 54.1% of its principal cancelled by FEMA. The total percentage of principal cancelled was 68.9%.

Legislative Proposal in FY2013 Homeland Security Appropriations

As reported out of the Senate Appropriations Committee, the Department of Homeland Security Appropriations Act, 2013 (S. 3216) includes a general provision that would alter existing procedures for cancelling SCDLs. If passed by Congress, this provision is likely to reopen the cancellation application and review process for most, if not all, remaining loans that have not been fully cancelled. Local governments would have until April 30, 2014, to submit a revised cancellation application, and FEMA would have to issue its determination and resolve all appeals by April 30, 2015. The corresponding DHS appropriations bill as passed by the House of Representatives, H.R. 5855, does not include a similar general provision on SCDLs. For all SCDLs not already cancelled in full, Section 560 of S. 3216 would direct the Administrator of FEMA to review the remaining un-cancelled or partially cancelled loans for possible further cancellation with newly prescribed procedures. There are 71 special loans, issued to 54 different local governments, that could be reviewed again by FEMA with new procedures. However, it is unclear how many of the 54 local governments would resubmit for possible cancellation. In total, there is approximately $324.1 million in principal and $52.3 million in interest from these loans that could be cancelled under these new procedures. This total includes approximately $28.2 million in principal and interest that local governments have already repaid to FEMA.

These repaid funds could be reimbursed by FEMA if a new cancellation review found that they should be cancelled under the new guidelines.

As discussed in the "Review Process for Loan Cancellation" section of the report, current cancellation procedures require FEMA to calculate the "operating budget deficit" of a local government over a three fiscal year period. This process depends critically on what is defined as "revenue," "operating budget," and a "disaster-related expense of a municipal operating character." The proposed change to the procedures would require FEMA to exclude revenues in the General Fund that are dedicated, by law, to be disbursed to special districts or other purposes.[117] FEMA would also be required to count disaster-related capital expenses not covered by insurance or other federal proceeds, debt-servicing expenses, and sick/leave pay as a "disaster-related expense of a municipal operating character." These issues are not currently covered specifically by law or in existing FEMA regulations, but it is presumed that these changes would revise the manner in which FEMA implements its existing cancellation process. Existing concerns over how FEMA previously implemented the SCDL cancellation application and review process are discussed in the "Controversy Surrounding SCDL Cancellation" section of the report.

If the proposed change is passed by Congress, FEMA *may* also consider a timeframe of three, five, or seven fiscal years after the date of the disaster declaration (which was August 29, 2005), instead of just the three fiscal year period prescribed under current law. It is unclear whether FEMA would exercise this option, and if so, if they would apply it consistently as a means of increasing or decreasing cancellation rates. In some circumstances, for instance, a seven fiscal year period may produce a lower cancellation amount than a five or three fiscal year review period, and vice versa. FEMA may choose to conduct an evaluation of each loan under each length of time period; under one of the two new options consistently (e.g., always review each application for a five year period), or simply continue to use the three fiscal year period standard. If FEMA chooses to review each loan under each time period, they would also then need to decide whether to choose the period that produces the least, median, or most amount of operating budget deficit (and thus the total amount of principal to be cancelled). Local governments would have until September 30, 2035 (roughly 30 years after Hurricane Katrina and the loans were issued), to repay the remaining amount of principal and interest not cancelled under the new procedures. This would extend the current maximum repayment period allowable by regulation, as discussed in the "Loan Repayment" section of the report, from 20 years to roughly 30 years.

It is impossible to predict the impact the new procedures would have on existing cancellation rates for SCDLs. In all likelihood, rates of cancellation will be higher under the new procedures, especially if FEMA chooses to evaluate each loan under the time period most beneficial to the applicant. However, the new cancellation amount will depend on the budgetary specifics of each local government and on decisions that FEMA would have to make to implement the law.

COMPARING TCDL AND SCDL CANCELLATION RATES

Part of the controversy surrounding the SCDLs is a general perception that the overall cancellation rates for traditional loans and special community disaster loans have differed, and specifically that SCDL cancellation rates were lower than those of TCDLs. Without conducting a full audit of the program and loan cancellation decisions, it is difficult to precisely evaluate this perception. By some statistical measures, TCDL rates for cancellation are lower than SCDL rates. By other measures, SCDLs have been forgiven at lower rates than TCDLs. **Table 8** and **Table 9** provide several measures for comparing the cancellation rates of TCDLs to SCDLs.

In summary, TCDLs had a lower percentage of loans fully cancelled or with some level of cancellation than SCDLs (33.9% and 46.4% versus 50.0% and 59.7%, respectively). On average, TCDLs also had lower dollar amounts of principal forgiven per loan than SCDLs (38.9% versus 54.1%). However, as a function of total dollar amount of principal cancelled in each loan category, TCDLs had a much higher cancellation rate than SCDLs (97.2% versus 68.9%).

Regardless of the data presented, the overall cancellation rates of each set of loans are not definitive evidence that the regulations and procedures used by FEMA were fairly or consistently applied, or that local governments in either set of loans received their "appropriate" amount of cancellation under the law. The disaster and financial circumstances surrounding each loan issued through the program are unique, and therefore aggregate assessments of the loan categories only provide a broad indication of how the program was managed. As a reminder, these tables only evaluate loans that had some amount of principal borrowed by the local governments, excluding a total of 39 loans that local governments did not borrow any amount from the approved principal.

                     Jared T. Brown

Table 8. Cancellation of Traditional and Special Community Disaster Loans
Cancellation Rates of TCDLs and SCDLs by Numeric Count of Loans

Loan Category	Government Type	Total Loans Eligible for Cancellation	Number of Loans Fully Canceled	Number of Loans Partially Canceled	Number of Loans with No Cancelled Balance	Percent of Loans Fully Cancelled	Percent of Loans with Some Cancellation[a]
Traditional	Territory[b]	3	1	2[c]	0	33.3%	100.0%
	Municipal Government[d]	36	14	4	18	28.9%	50.0%
	Special Districts, School Boards, and other[e]	17	4	1	12	23.5%	29.4%
	TCDL Sum:	56	19	7	30	TCDL Avg: 33.9%	46.4%
Special	Municipal Government	51	23	6	22	43.1%	56.9%
	Special Districts, School Boards, and other	78	35	13	30	44.9%	61.5%
	SCDL Sum:	129	58	19	52	SCDL Avg: 50.0%	59.7%
	All Loans Sum:	184	77	26	82	All Loans Avg: 41.8%	56.0%

Source: Calculations and categorization by CRS. Raw data provided by FEMA staff, and were current as of June 15, 2012.

Notes: Percentages rounded to the nearest tenth of a percent. Loans in appeal (17) are categorized with by their cancellation status as of June 15, 2012. Four of the appealed loans have already been partially cancelled, and the remaining 13 have had no amount cancelled. Some of these loans may ultimately be partially or fully cancelled, which would ultimately increase the percentage cancelled. There is one traditional loan that has not completed its review for cancellation, and is therefore are excluded from this table. This table does not include loans where zero dollars were borrowed or the loan was withdrawn by the choice of the local government.

a. Combines number of loans fully or partially cancelled as a percentage of the total loans in each category.

b. This category includes U.S. territories.

c. One U.S. Virgin Island loan was partially cancelled using the normal traditional regulations and the remainder was cancelled by legislation. See discussion in "Summary of Traditional Loan Cancellation." Since FEMA partially cancelled the loan using the normal regulatory process, the loan is categorized as partially cancelled.

d. This category includes general purpose municipal city, county, or parish governments.

e. This category includes government entities such as water and sewage boards, transportation special districts like maritime ports and airports, hospitals, sheriff's departments, and school districts/boards.

Table 9. Cancellation of Traditional and Special Community Disaster Loans

Cancellation Rates of TCDLs and SCDLs by Total Dollar Amount of Loans, in Millions

Loan Category	Government Type	Total Principal Cancelled	Total Principal Not Cancelled	Total Principal and Interesta Cancelled	Total Principal and Interesta Not Cancelled	Percent of Total Principal Cancelled	Percent of Total Principal and Interesta Cancelled	Average Percent of Principal Cancelled per Loanb	Average Percent of Principal and Interesta Cancelled per Loanb
Traditional	Territoryc	$135.8	$1.5	$200.9	$3.1	99.1%	98.5%	92.4%	89.7%
	Municipal Government	$41.7	$3.6	$54.5	$5.5	92.1%	90.8%	42.1%	43.1%
	Special Districts, School Boards, and other	$2.3	$0.1	$2.9	$0.1	95.8%	96.7%	25.9%	25.7%
	TCDL Sum:	$179.8	$5.2	$258.3	$8.8	TCDL Avg: 97.2%	96.7%	38.9%	39.4%
Special	Municipal Government	$315.5	$128.8	$349.7	$149.5	71.0%	70.1%	52.0%	52.0%
	Special Districts, School Boards, and other	$401.2	$195.3	$446.1	$226.8	67.3%	66.3%	55.5%	55.4%
	SCDL Sum:	$716.6	$324.1	$795.8	$376.4	SCDL Avg: 68.9%	67.9%	54.1%	54.1%
	All Loans Sum:	$896.5	$329.3	$1054.2	$385.1	All Loans Avg: 73.1%	73.2%	49.6%	49.7%

Source: Calculations and categorization by CRS. Raw data provided by FEMA staff, and were current as of June 15, 2012.

Notes: Dollars listed in millions. Dollar figures have been rounded to nearest tenth of a million. Percentages rounded to the nearest tenth of a percent. Totals may not add due to rounding error. See notes for Table 8 for additional explanations and definitions.

a. Amount of interest accrued per loan is subject to two factors, namely the interest rate applied to each loan and the period of time the loan is outstanding. Since these factors are not equal across individual loans or the categories of loans, presenting cancellation rates with interest amounts is potentially controversial because some loans may have relatively higher amounts of interest accrued by comparison to others.

b. These figures represent the average, across loan categories, of the percent of loan forgiveness per loan. In other words, if a loan was completely cancelled, its percentage of cancellation is 100%, and if it had no cancellation, its percentage is 0%. A partial loan forgiveness would range from 0 to 100%, such as 25% if a quarter of loan was cancelled. By averaging these percentages, this measure treats each loan equally, regardless of total dollar size of the loan. For example: Loan A had 25% of its principal cancelled. Loan B had 75% of its principal cancelled. The average percent cancelled per loan for those two loans is 50%, even if Loan A was for $100 million and Loan B was for $5 million.

c. The U.S. Virgin Island loan that was cancelled in part by legislative action was excluded from calculations. Exclusion was necessary because not enough information is available on the balances of principal and interest that were cancelled through traditional regulation processes versus the legislation. See discussion in "Summary of Traditional Loan Cancellation."

2008 COMMUNITY DISASTER LOANS

The 2008 calendar year was marred by numerous major disasters, notably Hurricanes Ike and Gustav that impacted parts of Texas and Louisiana and historic flooding that impacted wide swaths of Iowa.[118] In response to the disasters, the 110th Congress passed the Consolidated Security, Disaster Assistance, and Continuing Appropriations Act, 2009, which included the Disaster Relief and Recovery Supplemental Appropriations Act, 2008 (P.L. 110-329, henceforth Disaster Supplemental of 2008).[119] Through the Disaster Supplemental of 2008, the DADLP Account was appropriated $98.15 million for the CDL program, to subsidize gross obligations for the principal amount of direct loans not to exceed $100 million.[120]

The original appropriation in the Disaster Supplemental of 2008 did not create unique provisions for the administration of the loans, and the funds were available until expended. However, two subsequent laws passed in the 111th Congress established unique conditions for eligibility and loan size specific to disasters in the 2008 calendar year. The first law to create unique provisions for the 2008 CDLs was the American Recovery and Reinvestment Act of 2009 (P.L. 111-5, henceforth ARRA).[121] The second revision came in the Supplemental Appropriations Act, 2009 (P.L. 111-32, henceforth Supplemental of 2009). Neither ARRA nor the Supplemental of 2009 included any additional funds for the DADLP account. FEMA has not issued new regulations in response to the unique provisions of the 2008 loans.

Distinguishing Features from Traditional Loans

Eligibility

ARRA revised the 2008 CDLs in two significant ways. First, it created unique provisions for loan size that are discussed below. Second, it restricted these provisions only to applicants impacted by disasters in the calendar year of 2008. Therefore, although the appropriated amount of $98.15 million in the Disaster Supplemental of 2008 was designated as no year funds, the special provisions on the size of the loans available only applied to disasters in the 2008 calendar year.

The second revision to eligibility came in the Supplemental of 2009 (P.L. 111-32). Section 608 of the law provided an exemption for local governments in Texas that were impacted by Hurricane Ike.[122] The special provision changed the traditional program regulations on eligibility for the loans based on loss of revenue. Normally, FEMA uses the fiscal year of the disaster or the succeeding fiscal year as the base period for determining whether the loss in revenue is sufficient for need.[123] Since the Hurricane Ike disaster was declared on September 13, 2008, the years normally evaluated would have been FY2008 and FY2009. The law altered this by allowing FY2010 to be included as a base period for revenue loss evaluation for eligibility.[124] There were 14 different loans issued to 13 different governments under the Hurricane Ike declaration for Texas, but it is unknown how many times this provision was used in their application for a loan.

Number of Loans Issued to a Local Government

As with the SCDLs, FEMA waived the regulatory provision preventing only one loan being issued per disaster to a community.[125] This allowed eligible communities who received an original loan under the Disaster Supplemental of 2008 to apply for an additional loan under the special loan size provisions of ARRA. Only one local government received an additional loan under this exemption.

Loan Size

In ARRA, Congress altered the loan size for the 2008 CDLs by eliminating the $5 million cap and increasing the potential size of the loans from 25% to 50% of the prior year operating budget. A local government could only receive this larger percentage size and an amount more than $5 million if they suffered a projected loss of at least 25% in tax revenues. For the importance of the distinction between just tax revenues versus tax and other revenues, see the "Tax and Other" Revenue Clause" section of the report. The loan size provision in ARRA was identical to the changes made to loan size by the Emergency Supplemental of 2006 for Hurricane Katrina and the 2005 hurricane season (P.L. 109-234). Only one additional loan was issued to a government under this provision, allowing the local government to receive a loan above the 25% operating budget threshold. Several additional governments reached the $5 million cap (eight in total), but, since they did not meet the requirement of a 25% loss in tax revenues, the cap was not waived.

Summary of 2008 Loans

FEMA approved 24 loans to 23 separate local governments in 3 different states (5 each to Iowa and Louisiana, and 14 to Texas). Of these loans, 20 were used by 19 different governments, meaning some amount of the money was actually borrowed by a government. Thus far, local governments have borrowed approximately $49.7 million from the $60.6 million that was approved, which is $10.9 million less than their full eligible amount. **Table 10** summarizes by government type the amount of loans issued through the 2008 legislative exemptions.

Timeline for 2008 Community Disaster Loan Cancellation

Unlike the original laws passed for the SCDLs, neither ARRA nor the Supplemental of 2009 contained any unique provisions altering the ability of the 2008 CDLs to be cancelled under Section 417(c)(1) of the Stafford Act. Therefore, it is FEMA's stated intention to review the forgiveness of 2008 CDLs for using the "traditional" rules for loan forgiveness, not through the SCDL rules (44 C.F.R. §206.366 versus 44 C.F.R. §206.376). However, one might reasonably expect that the expanded definitions for terms such as "operating expenses" and "revenue" would be used for future forgiveness evaluations by FEMA. These new definitions were meant to provide more clarity than the traditional regulations.[126]

Table 10. Borrowed 2008 Loan Amounts by Government Type

Government Type	Number of Loans	Total $ Amount of Borrowed Principal, in Millions	Average $ Amount of Borrowed Principal per Loan, in Millions
Municipal Government[a]	10	$22.4	$2.2
Special Districts, School Boards, and other[b]	10	$27.3	$2.7
Total	**20**	**$49.7**	**$2.5**

Source: Calculations and categorization by CRS. Raw data provided by FEMA staff, and were current as of June 15, 2012.

Notes: This table does not include loans where no funds were borrowed or the loan was withdrawn by the choice of the local government (four loans excluded). Dollar figures have been rounded to nearest tenth of a million. Totals may not add due to rounding error.

a. This category includes general purpose municipal city, county, or parish governments.

b. This category includes government entities such as water and sewage boards, transportation special districts like maritime ports and airports, hospitals, sheriff's departments, and school districts/boards.

The 2008 CDLs recently became eligible for cancellation, as they reached the three-fiscal-year mark since the disaster on October 1, 2011. However, under the cancellation regulations, they each have until five years after the date of the promissory note to submit their application.[127] The earliest maturity date for these loans is January 26, 2014. There may be an incentive to apply early for cancellation because interest on the loans will continue to accrue over the next few years. If a portion of the loan is not cancelled, it may behoove local governments to pay back the remaining principal of loan to avoid this accrued interest. The average interest rate for these loans is 1.89%.

ISSUES FOR CONGRESS

As discussed in the "Frequency of Program Use" section of the report, the community disaster loan program is rarely used by local governments relative to other common disaster assistance programs. However, when the program has been used, it has provoked controversies—especially surrounding the authority to cancel the repayment of loans. In future appropriations and authorizations of the CDL program, Congress may consider some of the policy issues and options discussed succinctly below. Some of the options presented, if considered more thoroughly and enacted by Congress, may increase or decrease the disaster assistance provided through the program, and thus the overall use of the program by local governments.

It is beyond the scope of this report to fully discuss or evaluate the policy issues and options provided here. However, CRS is available to confidentially assist committees and Members of Congress and their staff as they analyze the potential policy implications of various proposals.

Considerations for Amending Components of the CDL Program

Congress could consider changing some of the core components of the program discussed throughout the report, namely the eligibility, use, size, and cancellation of the loans. Some of the many options for doing so are discussed briefly here.

Options Regarding Eligibility

As discussed in "Eligibility Criteria," there are only three eligibility conditions provided in the law authorizing the CDL program. First, local governments must have suffered losses "as a result of a major disaster." Second, a local government should have "suffer[ed] a substantial loss of tax and other revenues." Third, there "has to be a demonstrated need for financial assistance" for the loan to be issued.[128] FEMA has developed these three statutory conditions into a full set of regulations on eligibility, outlined in 44 C.F.R. §206.363. Congress could alter these legal and regulatory conditions to change the scope, use, and efficiency of the program.

Type of Eligible Government

Currently, program eligibility is restricted to local governments in declared disaster areas. As discussed in the report, the regulatory definition of local government allows a broad group of government entities to apply for loans.[129] From an implementation standpoint, it may be too difficult to craft a one-size-fits-all regulation for these local government types, especially when considering the different types of operating budgets, revenues, and expenditures encompassed by these governments. Additionally, though some standard methods exist for government accounting (namely those established by the Government Accounting Standards Board), each jurisdiction does not necessarily interpret these standards in the same way. In response to this implementation challenge, Congress could consider restricting loans only to a higher level of government jurisdiction. For instance, instead of all local governments being eligible for loans, loans could be offered exclusively to states that have been impacted by a disaster. The state could then disburse the loan proceeds to local governments as it deems appropriate with its own requirements and procedures. Or, loans could be made at the county or parish level, which is the level of distinction used to declare disasters through the Stafford Act. Allowing loans to be offered to states has been proposed in legislation before, though local governments were also eligible in that proposal.[130] In conjunction with this change, the size of loans may need to be increased to adjust to the larger jurisdiction level. Existing methods for assessing loan size (expected revenue loss calculations) could still be done at the higher jurisdiction level.

Conversely, Congress could consider restricting the type of local government that is eligible for a loan to a more specified set of government entities, such as only general purpose municipal governments. This would reduce some of the complexity from the program by eliminating special districts, school boards, and other types of local governments that often have unique revenue streams and budget requirements. However, these types of local governments often also provide essential government services not provided by general purpose municipal governments. Therefore, eliminating their program eligibility could reduce the potential benefits of the disaster assistance for the whole community.

Altering the jurisdiction level, or restricting the type of local governments, eligible for disaster loans could potentially improve the efficiency and transparency of the program. It may allow FEMA to establish a uniform set of regulations and procedures that are easier to implement, and easier for potential recipients to understand, in complex disaster situations. This could lead to less controversy and confusion regarding program use and its requirements, as well as reducing the administrative costs of the program. However, from the perspective of local governments currently eligible for the program, shifting the program to a higher level jurisdiction may lead to new challenges in that their county or state governments could set administrative guidelines for the loans that are more stringent or cumbersome than FEMA's existing procedures.

Additionally, the federal government may lose some level of transparency on how the disaster assistance is spent beyond the county or state level, whereas the current program has a level of oversight at the local government jurisdiction.

Threshold of Loss for Eligibility

Congress could also consider changing the eligibility criteria related to the statutory requirement that governments must have experienced a "substantial loss of tax and other revenues" to receive a loan.[131]

FEMA's major regulatory interpretation of this provision is that a government must indicate an expected 5% loss in revenue in order to receive a loan. There are also special provisions for loan size in the law that use expected revenue loss as a method for determining what loan size a local government is eligible to receive.[132] Congress could increase or decrease the amount of expected revenue loss required as a means of increasing or decreasing the use of the program in the future, respectively.

Further, Congress could make permanent the provisions of the Emergency Supplemental of 2006 that restricted loans of a certain size to governments that had experienced a 25% loss or more in exclusively tax revenues, but not all types of revenues.

By restricting eligibility to only governments that experience a significant loss of tax revenues, eligibility would also essentially be restricted to the types of local governments with tax revenue streams.

Elimination of "Substantial Loss" and "Demonstrated Need" Eligibility

It may be too difficult to develop and administer an objective set of eligibility requirements that adequately address when a local government has "lost" enough revenue to "need" disaster assistance. For instance, depending on the financial situation of a particular government and the amount of disaster impact, a local government may truly "need" assistance after only losing 2%-3% of their tourism revenue, while in other instances, a local government may only "need" assistance after losing 8%-9% of their total revenue stream.

Given this issue, Congress could consider changing the program to eliminate all eligibility requirements related to the local government having a "substantial loss" in revenue and a "demonstrated need" for assistance. Instead, Congress could allow loans to be offered to any local government in a declared disaster area. This would allow local governments to self-select for the loan program, making their own determination if they need or want a loan from the federal government to cover any amount of "loss" in revenues. In this situation, loan size could still be restricted to some formula determination made by the federal government, or the local government could be allowed to determine the size of the loan (likely within reasonable boundaries, or up to the amount they would be willing to provide in collateral). It is immediately unclear what the new level for demand would be for loans if these eligibility requirements were eliminated. However, this proposal could be combined with legislative revisions that either increase or decrease the interest rate or standards for loan cancellation to make the loans more or less attractive to local governments, respectively.

Options Regarding the Use of Loan Funds

As discussed in the "Use of Funds" and "Purpose of the Special Loans" sections of the report, there has been only one major policy change to how funds could be used in the history of the program. This occurred when SCDLs were restricted for "essential services" instead of a broader "government functions." FEMA has interpreted that neither "essential services" nor "government functions" include capital expenses or matches for other assistance programs.[133] As a means of increasing or decreasing the scope and use of the program, Congress may wish to consider amending the ways in which loan funds can be used by local governments.

To expand the scope, Congress could authorize the use of loan funds on capital projects that are not otherwise eligible for disaster assistance, namely through major disaster assistance programs under the Stafford Act.[134] These loans could help a local government more rapidly rebuild damaged infrastructure or build new facilities as a means of stimulating the disaster recovery process. As a result, these capital projects may, arguably, help restore the government's impacted sources of tax revenues from businesses and properties. However, allowing loan proceeds to be used for capital projects may expand the original intention of the program of being exclusively focused on government "services" or "functions." Because of the high cost of many capital projects, it is possible that the demand for loans, and their overall size, would increase in this proposal.

Conversely, Congress may wish to restrict the use of loan funds to only certain types of government functions—such as law enforcement, education, or public health—as a means of ensuring that federal assistance is only spent on the certain local government functions. However, many individualized programs already exist that target specific government functions, such as relief funds for transportation infrastructure damage, and the loan program may become redundant to these programs depending on the functional area.

Options Regarding the Dollar Size of Loans

Currently, loan size is capped by both law and regulatory provisions, as discussed in the different "Loan Size" sections of the report. In various iterations of the program, loan sizes have been capped in law by a strict dollar amount ($5 million) and by a percentage of the operating budget of the recipient local government (25% and 50% in different eligible conditions). Additionally, in regulations, FEMA has restricted the size of loans to a formula used to determine the "need" of the local government, essentially the sum of the projected revenue loss and projected unreimbursed disaster-related expenses of the recipient local government.[135]

As a means of controlling the overall cost of the program and the amount of assistance offered to the local government, Congress could consider revising any one of these methods of determining loan size. Strict dollar caps per local government can be useful because the maximum cost of a loan is predictable, regardless of disaster or local government factors. However, Congress may view a total dollar limit as relatively inequitable to local governments of a larger size, because the relative amount of assistance they receive is less than smaller governments as a percentage of their budgets or revenue streams.

Another alternative would be to formally establish a version of FEMA's formula for calculating "need" in program statute, and remove all other caps on the size of the loans. Congress could alter the existing need calculation in various ways, including specifically defining what constitutes an "unreimbursed disaster-related expense" or restricting the formula only to projected revenue losses of some kind. Congress could also consider setting the loan size to the sum of itemized predicted expenditures by the local government. Instead of using a specific numeric formula to determine need, Congress could replace the need formula with the independent assessment of expert accountants that could determine the "need" of the local government through some objective analysis of their financial condition. An independent analysis may capture factors such as the scope of disaster devastation specific to the event more readily than existing methods. However, any independent analysis, though theoretically impartial, may be critiqued for being inequitably applied in some disaster scenarios.

Options Regarding the Cancellation of Loans

Historic cancellation rates for the program are provided in **Table 8** and **Table 9**. Congress may wish to evaluate different proposals either increasing or decreasing the likelihood of loans being cancelled, as a means of increasing or decreasing the assistance provided to local governments.

Any proposal increasing the likelihood of a loan being forgiven, or of making forgiveness automatic (and thereby converting the loan program to a grant program), is likely to increase the federal cost and use of the program.

Eliminating the Possibility of Cancellation

Congress may wish to eliminate the possibility of loans being cancelled in whole or in part, as was done under the initial legislation appropriating for the SCDLs. There are several potential benefits to eliminating the loan cancellation provision. First, not allowing cancellation could reduce, or potentially eliminate altogether, the cost of the program to the taxpayer. Depending on the interest and default rates on the loans, the program could be close to if not completely budget neutral (or even provide revenue for the federal government if interest rates were high). Second, not allowing cancellation could reduce the administrative costs of the program for both the federal and local government, as it would rid the need for costly reviews and appeals on old loans. Finally, structuring it as a loan program without cancellation could reduce the hypothetical use of local governments who apply for and receive the loans without a true "need" for the assistance after a disaster. Of course, this also means that the program provides less true "assistance" to the local governments following a disaster. Communities that are struggling with continued revenue shortfalls may be unduly burdened by the loan repayment.

Instead of completely restricting cancellation, Congress may wish to only allow loans to be cancelled up to a certain percentage of the initial principal (for instance, 50% of the principal). This option could still provide some benefit to the local governments that are financially struggling years after a disaster, while constraining the overall costs of the program to federal taxpayers.

Revision of Standards for Cancellation

Instead of eliminating the cancellation of loans altogether, Congress could consider changing the metrics used to determine whether or not a loan should be forgiven by FEMA. In theory, the current statutory standard for cancellation—whether revenues of a local government are "insufficient" to meet their operating budget after three fiscal years[136]—is intended to serve as proxy measure for the "need" or fiscal "health" of local government as it continues its disaster recovery. Following a disaster, local governments can be faced with a difficult choice of either continuing their pre-disaster spending levels in order to stimulate recovery in hopes of returning their community "back to normal," or cutting their budgets in response to a continued shortfall in revenues. Some have criticized the program for "penalizing" those local governments that attempt to restrain their operating costs following a disaster in a fiscally prudent manner in response to this dilemma.[137] One alternative metric Congress could consider is to only evaluate actual loss of revenue of a local government during the three-year fiscal period based on pre-disaster standards, regardless of whether or not revenues were sufficient to meet the operating budget of the local government post-disaster.[138] This metric could be simpler to implement, as it would remove FEMA from the position of evaluating the budgetary decisions and conditions of the local government post-disaster. However, such a process opens the possibility that a local government that truly did not "need" (by some subjective measure) the loan funds to provide government services following a disaster could receive the superfluous assistance from the federal government. Another alternative would be to create a process that does not rely exclusively on objective accounting calculations of revenue, budgets, and expenditures, and instead allows an independent panel of experts to evaluate the disaster recovery progress of the local government through some objective and subjective analysis. If the expert panel determined a local government was still struggling to provide government services due to the disaster, the panel could recommend forgiveness of the loan to FEMA. Of course, any subjective decision-making process for cancellation is prone to criticisms of being inequitably applied, of lacking transparency, or of being motivated and manipulated by political considerations.

Option to Convert to a Grant Program

Instead of not allowing any cancellation, Congress may wish to fully cancel all "loans" by creating a grant program instead of a loan program. When the CDL program was first established in 1974, it actually replaced a grant program created in 1970 with similar eligibility requirements. The grant program was authorized in Section 261 of the Disaster Relief Act of 1970 (P.L. 91-606, 84 Stat. 1744), and allowed the President to make grants to any local government which, as the result of a major disaster, had suffered a substantial loss of property tax revenue.

Similar to the structure of the 1970 grant program, Congress may wish to replace the loan program with grants that have a set size based on lost revenue. One option would be to replicate the procedures that FEMA follows now to determine loan size and use them to determine grant size instead. With a grant program, immediate revenue relief could be provided to local jurisdictions in a disaster area without saddling them with additional debt. With no possibility of interest or principal repayments, a grant program could cost more per dollar of aid delivered than a loan program. In addition, a grant program is likely be used by more jurisdictions than a loan program and could therefore be considerably more expensive for federal taxpayers.

Conversion to a grant program could also allow more local governments to receive assistance. Some local governments are barred from the current loan program because they legally cannot incur debt to fund government operations, or they are not allowed to borrow directly from the federal government.[139] Similarly, many local governments are required to maintain a balanced budget, and therefore would not have a cumulative operating deficit after the three fiscal year period. FEMA has noted in rulemakings that balanced budget procedures prevent these governments from having their loans cancelled.[140]

Option to Convert to a Hybrid Loan and Grant Program

Congress could also consider transforming the program into a hybrid grant and loan program. In this circumstance, a grant could be offered between certain thresholds of revenue loss (for instance, between 5% and 10% of expected revenue loss), and a loan could be issued for the remainder of the need of the local government. The loan could be offered with or without the possibility of being cancelled in conjunction with the grant. A hybrid program that can offer both grants and loans has some precedent. For instance, the Rural Communities Facilities Program managed by USDA offers both grants and loans for the same purpose.[141] However, the various complexities of managing a hybrid program, as well as financing it through appropriations, are certainly worth considering when evaluating this proposal. Converting the disaster loan program into either a full or hybrid grant program may also involve modifying the current appropriation account, the Disaster Assistance Direct Loan Program account. If the disaster loan program became a full grant program, Congress could consider funding the program through the Disaster Relief Fund, as is practice for most other Stafford Act programs.

Considerations for Eliminating the CDL Program

The core purpose of the CDL program is to provide financial assistance to local governments that are having difficulty providing government services because of a loss in tax or other revenue. In the future, Congress may reevaluate this core purpose, irrespective of the method and means of providing the assistance. Congress may no longer consider it in the interest of the federal government to aid local governments experiencing revenue loss after a disaster, especially since there are a wide array of more targeted disaster assistance programs available to local governments.[142] Or, Congress may conclude that supporting the maintenance of local government operating budgets through the loan program violates the principles of federalism upon which most disaster assistance programs are based. If Congress wishes to end the CDL program, it could do so by ending appropriations to the DALDP account for the purposes of the program, or by eliminating its authorization in the Stafford Act.

As an alternative to direct loan assistance through the CDL program, Congress could consider alternative programs designed to restore certain types of local government revenues following a disaster. For instance, some local government financial stress related to revenue loss could be alleviated by programs that promote the return of businesses or tourism to the region. Existing programs, such as the Small Business Administration's Disaster Loan Program,[143] could be reviewed by Congress to ensure they adequately address the needs of private businesses in this regard.

End Notes

[1] The number of employees of the New Orleans city government decreased by approximately 34% from 2005 (pre-hurricane) to 2009 (post-hurricane). Calculation drawn from the 2005 and 2009 versions of the U.S. Census Bureau, *Annual Survey of Public Employment and Payroll*, Washington, D.C., http://www.census.gov/govs/apes/.

[2] For a discussion on how states and local governments can and cannot incur debt, see CRS Report R41735, *State and Local Government Debt: An Analysis*, by Steven Maguire.

[3] U.S. Government Accountability Office, *Disaster Recovery: Past Experiences Offer Insights for Recovering from Hurricanes Ike and Gustav and Other Recent Natural Disasters*, GA0-08-1120, September 2008, p. 17, http://www.gao.gov/products/GAO-08-1120.

[4] U.S. Congress, Senate Committee on Public Works, *Disaster Relief Act Amendments of 1974*, Report to Accompany S. 3062, 93rd Cong., 2nd sess., April 9, 1974, S.Rept. 93-778, p. 9.

[5] As defined in 42 U.S.C. §5122 of the Stafford Act, the definition of "local government" is:
(A) a county, municipality, city, town, township, local public authority, school district, special district, intrastate district, council of governments (regardless of whether the council of governments is incorporated as a nonprofit corporation under state law), regional or interstate government entity, or agency or instrumentality of a local government;
(B) an Indian tribe or authorized tribal organization, or Alaska Native village or organization; and
(C) a rural community, unincorporated town or village, or other public entity, for which an application for assistance is made by a state or political subdivision of a state.

[6] P.L. 93-288, as amended; 42 U.S.C. 5121 et seq. The 1974 Act was renamed the Stafford Act by the Disaster Relief and Emergency Assistance Amendments of 1988, P.L. 100-707, Section 102. Prior to Act in 1988, the CDL Program was codified as Section 414, not Section 417.

[7] 42 U.S.C. §5184(a).

[8] 6 U.S.C. §314(a)(8).

[9] 42 U.S.C. §5184.

[10] The regulations for TCDLs are found in 44 C.F.R. §206.360-367. The regulations for the SCDLs are found in 44 C.F.R. §206.370-377.

[11] The 2008 CDLs are governed by the regulations applicable to traditional CDLs, found in 44 C.F.R. §206.360, with certain provisions waived or altered by FEMA to account for the special legislation.

[12] Unless otherwise specified in the report, the data presented is calculated and categorized by the author from raw data on loans provided by FEMA. The information was current as of June 15, 2012.

[13] For more on this debate, see CRS Report R42352, *An Examination of Federal Disaster Relief Under the Budget Control Act*, by Bruce R. Lindsay, William L. Painter, and Francis X. McCarthy.

[14] For instance, in every disaster declaration, either Individual or Public Assistance is made available through the Stafford Act. For more on these programs, see CRS Report RL33053, *Federal Stafford Act Disaster Assistance: Presidential Declarations, Eligible Activities, and Funding*, by Francis X. McCarthy.

[15] In most appropriation bills, the program will be identified either by this account or as Section 417 of the Stafford Act.

[16] 42 U.S.C. §5162.

[17] The Omnibus Budget Reconciliation Act of 1990, P.L. 101-508, added Title V to the Congressional Budget Act. Title V is also known as the Federal Credit Reform Act of 1990.

[18] For more on how credit subsidy rates are calculated and their purpose, see CRS Report RL30346, *Federal Credit Reform: Implementation of the Changed Budgetary Treatment of Direct Loans and Loan Guarantees*, by James M. Bickley, especially Appendix B,

Budgetary Treatment of a Hypothetical Direct Loan. The Office of Management and Budget also produces subsidy rate estimates.

[19] P.L. 110-329. See 122 Stat 3592 for provision.

[20] For OMB's most recent estimates of the CDL program, see U.S. Executive Office of the President, Office of Management and Budget, *Budget of the United States Government Fiscal Year 2013, Federal Credit Supplement*(Washington: February 2012), Table 7, p. 44.

[21] Section 417(a) of the Stafford Act.

[22] 44 C.F.R. §206.363(a)(1). For more on how disaster declarations are made under the Stafford Act, see CRS Report RL33053, *Federal Stafford Act Disaster Assistance: Presidential Declarations, Eligible Activities, and Funding*, by Francis X. McCarthy.

[23] 42 U.S.C. §5122. See footnote 5.

[24] See discussion in Federal Emergency Management Agency, "Disaster Assistance; Community Disaster Loans," 53 *Federal Register* 12681, April 18, 1988.

[25] 44 C.F.R. §206.361(d).

[26] Section 417(c)(2) of the Stafford Act.

[27] 44 C.F.R. §206.363(b)(2).

[28] 44 C.F.R. §206.363(3)(i-ix). One of the factors listed in the regulation, (vi), repeats the same provision in (iv).

[29] Specifically, provisions in 44 C.F.R. §206.363(3)(i), (ii), (iii), (v), and (ix).

[30] The application process is outlined in 44 C.F.R. §206.364. For the notice of information collection detailing the forms in the application, see Department of Homeland Security, "Agency Information Collection Activities: Submission for OMB Review; Comment Request," 74 *Federal Register* 17971, April 20, 2009.

[31] 44 C.F.R. §206.364(a). The GAR is "The person empowered by the Governor to execute, on behalf of the state, all necessary documents for disaster assistance," as defined in 44 C.F.R. §206.2(a)(13).

[32] See FEMA Form 090-0-1, "Certification of Eligibility for Community Disaster Loans," part of the overall "Application for Community Disaster Loan Program," OMB No. 1660-0083.

[33] 44 C.F.R. §206.364(b).

[34] 44 C.F.R. §206.364(d)(2).

[35] Federal Emergency Management Agency, "Disaster Assistance; Community Disaster Loans," 53 *Federal Register* 12681, April 18, 1988.

[36] Telephone conversation with FEMA program staff for the Community Disaster Loan Program, September 22, 2011. Typically, the collateral pledge by local governments is future revenues. An example of a promissory note and pledge of revenues for a special community disaster loan can be found at http://www.lafourchegov.org/ORDINANCES/2006/3728.pdf.

[37] If evidence of actual revenue losses were required, the distribution of the loan would likely be delayed by a year or more.

[38] 44 C.F.R. §206.364(b)(4).

[39] 44 C.F.R. §206.364(c)(2).

[40] In-person conversation with FEMA program staff for the Community Disaster Loan Program, August 3, 2011, and telephone conversation with FEMA program staff for the Community Disaster Loan Program, September 22, 2011.

[41] Section 417(b)(1) of the Stafford Act.

[42] Section 207(5) of P.L. 106-390, the Disaster Mitigation Act of 2000.

[43] U.S. Congress, House Committee on Transportation and Infrastructure, *Disaster Mitigation and Cost Reduction Act of 1999*, Report to Accompany H.R. 707, 106th Cong., 1st sess., March 3, 1999, H.Rept. 106-40 (Washington: GPO,2000), p. 2.

[44] See H.R. 5523 of the 107th Congress. The bill was introduced by Rep. Carolyn Maloney of New York's 14th district, and co-sponsored by 16 other Members of the New York delegation. The bill would have eliminated the $5 million cap on the size of the loans, allowed states to also receive loans, and automatically cancelled any loans issued following a terrorist attack. According to a press release by Rep. Maloney, the bill also had the

support of New York's Senators Charles Schumer and Hillary Clinton. See Office of U.S. Representative Carolyn B. Maloney, "New York Lawmakers Seek Relief for Lost Tax Revenue Post 9/11," press release, October 2, 2002, http://maloney.house.gov/index.php?Itemid=61&id=551&option=com_content&task=view.

[45] Though the loans in 2007 were issued with a $5 million cap in place, none of the loans approached that limit due to other regulatory restrictions.

[46] 44 C.F.R. §206.364(d)(1)(i).

[47] 44 C.F.R. §206.361(c).

[48] Many state and local governments are not authorized, by local statute or constitutions, to issue long-term debt to cover operating expenses.

[49] 44 C.F.R. §206.365(a).

[50] Telephone conversation with FEMA program staff for the Community Disaster Loan Program, September 22, 2011.

[51] Section 417(a) of the Stafford Act.

[52] Department of Homeland Security, "Special Community Disaster Loans Program," 70 Federal Register 60444, October 18, 2005.

[53] 44 C.F.R. §206.361(f).

[54] Section 417(c)(1) of the Stafford Act.

[55] Section 414 of P.L. 93-288.

[56] 44 C.F.R. §206.366(d)(1).

[57] 44 C.F.R. §206.366(c). As previously noted in footnote 31, the GAR is "The person empowered by the Governor to execute, on behalf of the state, all necessary documents for disaster assistance," as defined in 44 C.F.R. §206.2(a)(13).

[58] 44 C.F.R. §206.367(b)(1).

[59] 44 C.F.R. §206.366(c)(2).

[60] See Department of Homeland Security, "Special Community Disaster Loans Program," 74 Federal Register 15229, April 3, 2009; Department of Homeland Security, "Special Community Disaster Loans Program," 75 Federal Register 2800, January 19, 2010; and "Special Community Disaster Loan (SCDL): Cancellation Process and Procedures," PowerPoint presentation available to local governments, February 2010.

[61] 44 C.F.R. §206.366(c)(4).

[62] 44 C.F.R. §206.366(a)(1).

[63] 44 C.F.R. §206.366(a)(5).

[64] 44 C.F.R. §206.366(a)(2).

[65] Ibid.

[66] 44 C.F.R. §206.366(b)(1). Examples of UBREs are provided in 44 C.F.R. §206.366(b)(3)(i-iv).

[67] 44 C.F.R. §206.366(b)(2).

[68] 44 C.F.R §206.367(c).

[69] 44 C.F.R. §206.361(e) and 206.367(c).

[70] 44 C.F.R. §206.364(d)(2).

[71] 114 Stat. 938.

[72] 115 Stat. 433-434. Also see H.Rept. 107-103 for a full explanation on how the transferred funds cancelled the remaining balance of the loan.

[73] For more on how credit subsidy rates are calculated and their purpose, see CRS Report RL30346, *Federal Credit Reform: Implementation of the Changed Budgetary Treatment of Direct Loans and Loan Guarantees*, by James M. Bickley.

[74] Section 2(a) of P.L. 109-88.

[75] 44 C.F.R. §206.361(a).

[76] Department of Homeland Security, "Special Community Disaster Loans Program," 70 *Federal Register* 60444, October 18, 2005.

[77] Department of Homeland Security, "Special Community Disaster Loans Program," 74 *Federal Register* 15229, April 3, 2009.

[78] Department of Homeland Security, "Special Community Disaster Loans Program," 75 *Federal Register* 2803, January 19, 2010.

[79] 119 Stat. 2061.

[80] After Hurricane Katrina, major hurricane disaster declarations were issued for counties in Alabama, Louisiana, Mississippi, and Florida—numbers 1605, 1603, 1604, and 1605, respectively.

[81] During the entire 2005 hurricane season, major hurricane disaster declarations were issued for various hurricanes/tropical storms for counties in Alabama, Florida, Louisiana, Mississippi, North Carolina, and Texas. Both North Carolina and Texas had disaster declarations that fell in the 2005 hurricane season, but in FY2006—declarations 1608 and 1609, respectively. A full searchable database of disaster declarations is available on the FEMA website at http://www.fema.gov/news/disaster.fema.

[82] 44 C.F.R. §206.371(d).

[83] After Hurricane Rita, major disaster declarations were issued for counties in both Louisiana and Texas—numbers 1607 and 1606, respectively.

[84] See the removal of the single loan provision in the difference between 44 C.F.R. §206.361(d) and 44 C.F.R. §206.371(d).

[85] Department of Homeland Security, "Special Community Disaster Loans Program," 75 *Federal Register* 2803, January 19, 2010.

[86] 44 C.F.R. 206.374(d)(1)(i).

[87] Department of Homeland Security, "Special Community Disaster Loans Program," 74 *Federal Register* 15229, April 3, 2009.

[88] 120 Stat. 459.

[89] Fore example, see the statement of Representative Richard Baker, *Congressional Record*, daily edition, vol. 151, no.130, October 7, 2005, p. H8796.

[90] 44 C.F.R. §206.371(c).

[91] Interest rates varied slightly based upon the rates for five-year maturities on the date a promissory note was executed for each loan. The lowest interest rate for an SCDL was 2.38%, the highest was 3.12%.

[92] Section 2(a) of P.L. 109-88.

[93] Statements by Senator Hillary Clinton on Relief for the Gulf Coast and by Senator Bill Frist, *Congressional Record*, daily edition, vol. 151, no. 130, October 7, 2005, p. S11280 and p. S11282, respectively.

[94] Statements regarding the Community Disaster Loan Act of 2005, *Congressional Record*, daily edition, vol. 151, no. 130, October 7, 2005, pp. S11279-S11285 and H8794-H8796.

[95] S. 1872 and H.R. 4117 in the 109th Congress.

[96] See S. 87, H.R. 680, and H.R. 1144.

[97] See S. 253.

[98] See Section 4502 of P.L. 110-28, 121 Stat. 156.

[99] Department of Homeland Security, "Special Community Disaster Loans Program" 70 *Federal Register* 60443, October 18, 2005.

[100] Jonathan Tilove, "FEMA Loan Forgiveness in the Works: After Visiting La. and Gulf Coast, Napolitano Proposes Rule Change," *The Times-Picayune*, March 31, 2009. http://www.nola.com/news/?/base/news-1/1238478603239340.xml&coll=1.

[101] In addition to several administrative changes in other sections that were made to reference the possibility of loan cancellation, the proposed rule copied 44 C.F.R. §206.366 and added it as a section in the SCDL regulations, at 44 C.F.R. §206.376.

[102] Department of Homeland Security, "Special Community Disaster Loans Program," 74 *Federal Register* 15231, April 3, 2009.

[103] Department of Homeland Security, "Special Community Disaster Loans Program," 75 *Federal Register* 2802, January 19, 2010.

[104] These included comments from U.S. Senators Mary Landrieu and David Vitter, and Representatives Rodney Alexander, Charles Boustany, Bill Cassidy, Ahn "Joseph" Cao,

John Fleming, Charlie Melacon, and Steve Scalise. These current and former Members collectively requested that FEMA "swiftly finalize Community Loan Disaster (CDL) forgiveness regulations and expedite the application and review process... and take into full consideration the points made by all the CDL recipients when finalizing the regulations." For a full list of the comments FEMA received on the proposed rule, see regulations Docket ID FEMA- 2005-0051 at http://www.regulations.gov/#!docketDetail;dct=PS;rpp=100;po=0; D=FEMA-2005-0051.

[105] These changes are explained in full in Department of Homeland Security, "Special Community Disaster Loans Program," 75 *Federal Register* 2802 -2803, January 19, 2010.

[106] Department of Homeland Security, "Special Community Disaster Loans Program," 75 *Federal Register* 2804, January 19, 2010.

[107] See 44 C.F.R. §206.376(2) and (4).

[108] Department of Homeland Security, "Special Community Disaster Loans Program," 75 *Federal Register* 2807, January 19, 2010.

[109] See "Review Process for Loan Cancellation."

[110] For example, the decision not to cancel one large loan made to the U.S. Virgin Islands was repeatedly appealed until it was ultimately cancelled approximately four years after its initial maturity date. For background, see Michelle Dominique, "FEMA Forgives $185 Million Hurricane Marilyn Loan," *St. Croix Source*, October 26, 2004, http://stcroixsource. com/content/news/local-news/2004/10/26/fema-forgives-185-million-hurricane-marilyn-loan-0.

[111] Letter from Mary Landrieu, United States Senator, to Eric Holder, Attorney General of the United States, November 5, 2010. A carbon copy of this letter was also sent to Craig Fugate, Administrator of the Federal Emergency Management Agency, and Janet Napolitano, Secretary of the Department of Homeland Security.

[112] Letter from Daryl G. Purpera, CPA, CFE, Louisiana Legislative Auditor, May 27, 2011, to a general list of local government recipients in Louisiana that received SCDLs. Among other items, the auditor claimed it was unclear how FEMA treated depreciation expenses, compensated absences, revenues for Internal Service Funds, and operating expenditures paid from a capital funds account.

[113] U.S. Senator David Vitter, "Vitter, La. Parish Officials Write VP Joe Biden, Ask Him to Follow Through on Disaster Loan Commitment," press release, May 23, 2011, http://www. vitter.senate.gov/public/index.cfm?FuseAction=PressRoom.PressReleases&ContentRecord _id=1e427fc5-f18e-f1b2-ef2c-bfe1fad2174d&Region_id=&Issue_id= 473e7dcc-b51e-2d9f-6091-f6d24f2bfde1.

[114] Bruce Nolan, "Joe Biden Brings Good News for Local Governments' Financial Recovery from Hurricane Katrina," *The Times-Picayune*, January 15, 2010, http://www.nola.com/ politics/index.ssf/2010/01/ joe_biden_brings_good_news_for.html.

[115] Jonathan Tilove, "Louisiana Officials Ask Federal Government to Forgive Disaster Loans," *The Times-Picayune*, May 23, 2011, http://www.nola.com/politics/index.ssf/2011/05/ louisiana_officials_ask_federa.html.

[116] For example, Representative Cedric Richmond is quoted by local media as saying, "We suffered one of the worst natural disasters in recent history and were assured by Vice President Biden himself, during his New Orleans speech in January 2010, that area parishes would receive loan forgiveness." See Bob Ross, "Jefferson Parish Appealing to FEMA to Get $55 Million in Disaster Loans Forgiven," *The Times-Picayune*, May 18, 2011, http:// www.nola.com/politics/index.ssf/2011/05/jefferson_parish_appealing_to.html.

[117] For instance, if a general purpose municipal city collects fees on behalf of a public utility that it has to immediately, by law, disburse to that utility, those would not be considered "revenue" in calculating the revenues of that general purpose municipal city.

[118] For a full list of major disaster declarations during 2008, see http://www.fema.gov/news/ disasters.fema?year=2008.

[119] The Disaster Relief and Recovery Supplemental Appropriations Act, 2008, was Division B of the Consolidated Security, Disaster Assistance, and Continuing Appropriations Act, 2009.

[120] P.L. 110-329. See 122 Stat. 3592 for provision.

[121] See 123 Stat. 164 for the provision in ARRA.

[122] Disaster declaration FEMA–1791–DR.

[123] 44 C.F.R. §206.363(b)(2).

[124] As with ARRA, no explanation for this provision was provided in congressional documentation.

[125] 44 C.F.R. §206.361(d).

[126] In-person conversation with FEMA program staff for the Community Disaster Loan Program, August 3, 2011.

[127] 44 C.F.R. §206.366(c).

[128] Section 417(a) of the Stafford Act.

[129] See 42 U.S.C. §5122 and footnote 5 for the definition of a local government.

[130] See, for example, H.R. 5523 of the 107th Congress.

[131] Section 417(a) of the Stafford Act.

[132] Section 417(b)(2) of the Stafford Act.

[133] 44 C.F.R. §206.366(c) and 44 C.F.R. §206.376(c).

[134] For more on these programs, see CRS Report RL33053, *Federal Stafford Act Disaster Assistance: Presidential Declarations, Eligible Activities, and Funding*, by Francis X. McCarthy.

[135] See 44 C.F.R. §206.364(d)(1)(i).

[136] See Section 417(c)(1) of the Stafford Act and discussion in "Cancellation of Traditional Loans."

[137] For instance, see the Testimony of Doris Voitier, Superintendent, St. Bernard Parish Schools. U.S. Congress, Senate Committee on Homeland Security and Governmental Affairs, Ad Hoc Subcommittee on Disaster Recovery, *Five Years Later: Lessons Learned, Progress Made, and Work Remaining from Hurricane Katrina*, Field Hearing in Chalmette, LA, 111th Cong., 2nd sess., August 26, 2010, S. Hrg. 111-1007 (Washington: GPO, 2011), p. 26. Also, see comments made by Sen. David Vitter and others in Office of U.S. Senator David Vitter, "Vitter, La. Parish Officials Write VP Joe Biden, Ask Him to Follow Through on Disaster Loan Commitment," press release, May 23, 2011, http://www.vitter.senate.gov/public/index.cfm?FuseAction=PressRoom.PressReleases&ContentRecord_id=1e427fc5-f18e-f1b2-ef2c-bfe1fad2174d&Region_id=&Issue_id=473e7dcc-b51e-2d9f-6091-f6d24f2bfde1.

[138] This metric is one of many suggested by commenters to FEMA's final rule on the SCDL program. For more, see Department of Homeland Security, "Special Community Disaster Loans Program," 75 *Federal Register* 2807-2810, January 19, 2010.

[139] For more on this, see CRS Report R41735, *State and Local Government Debt: An Analysis*, by Steven Maguire.

[140] Department of Homeland Security, "Special Community Disaster Loans Program," 75 *Federal Register* 2809, January 19, 2010.

[141] For more, see CRS Report RL31837, *An Overview of USDA Rural Development Programs*, by Tadlock Cowan.

[142] For an abbreviated list of federal assistance programs, see CRS Report RL31734, *Federal Disaster Recovery Programs: Brief Summaries*, by Carolyn V. Torsell.

[143] For more on this program, see CRS Report R41309, *The SBA Disaster Loan Program: Overview and Possible Issues for Congress*, by Bruce R. Lindsay.

In: The Community Disaster Loan Program
Editor: Felix P. Cardano

ISBN: 978-1-62417-643-2
© 2013 Nova Science Publishers, Inc.

Chapter 2

AN EXAMINATION OF FEDERAL DISASTER RELIEF UNDER THE BUDGET CONTROL ACT[*]

*Bruce R. Lindsay, William L. Painter
and Francis X. McCarthy*

SUMMARY

On August 2, 2011, the President signed into law the Budget Control Act of 2011 (BCA, P.L. 112- 25), which included a number of budget-controlling mechanisms. As part of the legislation, caps were placed on discretionary spending for the next ten years, beginning with FY2012. If these caps are exceeded, an automatic rescission—known as sequestration—takes place across most discretionary budget accounts to reduce the effective level of spending to the level of the cap. Additionally, special accommodations were made in the BCA to address the unpredictable nature of the need for disaster assistance while attempting to impose discipline on the amount spent by the federal government on disasters.

The first section of this report addresses the traditional funding for major disaster declarations, both through annual requested amounts and through supplemental appropriations to meet greater, unanticipated costs.

---

[*] This is an edited, reformatted and augmented version of a Congressional Research Service publication, CRS Report for Congress R42352, prepared for Members and Committees of Congress, from www.crs.gov, dated February 10, 2012.

This section also explains the workings of the President's Disaster Relief Fund, a "no-year" fund that finances spending under the Robert T. Stafford Disaster Relief and Emergency Assistance Act (P.L. 93-288).

Next, this report provides a basic overview of how disaster assistance is appropriated, what factors affect how much the federal government spends on disasters, how disaster relief is impacted by the BCA, and what the policy implications are for disaster assistance going forward under the constraints of the BCA. Included in this review are discussions of disaster spending, how it is calculated under the BCA, and potential costs that may be excluded under that calculation. The report also touches on the increasing number of disaster declarations and both the possible causes and likely cost implications of a greater number of declarations.

INTRODUCTION

Concern over the size of the federal budget deficit and the national debt has brought congressional attention, as an ancillary issue, to both the amount of funding the federal government provides to states and localities for disaster assistance and the processes the federal government uses to provide that assistance. The amount of funding provided by Congress for declared emergencies and major disasters has grown considerably in the past decade, driven principally by the hurricane season of 2005.[1] Although funds have been reallocated at times from one account to another to provide for disaster-related assistance, disaster relief funding has historically not been fully offset. Disaster relief funding has usually not been tightly constrained, either, with supplemental spending bills often funding these activities outside the allocations of discretionary budget authority and outlays associated with budget resolutions. Two potential methods for reducing the impact of disaster assistance spending on the federal budget are (1) the use of offsets and (2) placing controls on the level of allowable spending.

Among its provisions, the Budget Control Act (P.L. 112-25, hereafter the BCA) provides a mechanism designed, arguably, to limit spending on major disasters declared under the Robert T. Stafford Disaster Relief and Emergency Assistance Act (P.L. 93-288, hereafter the Stafford Act), thereby limiting the impact of such spending on the budget deficit.[2] This report reviews current disaster funding practices and examines potential issues presented to disaster funding under the BCA mechanism. The report also discusses how OMB calculates the "allowable adjustment" to discretionary spending caps for disaster relief pursuant to the BCA, and the potential policy implications that may result from that process.

An Overview of Disaster Spending: Federal Disaster Assistance and the Stafford Act

The Stafford Act authorizes the President to declare a major disaster in response to a governor's request for federal assistance.[3] The declaration enables federal agencies to provide assistance to state and local governments overwhelmed by the incident. While the majority of federal assistance for major disasters to states and localities is provided through the Federal Emergency Management Agency (FEMA), other federal agencies and offices also provide assistance once a major disaster has been declared. These include the U.S. Army Corps of Engineering, the Department of Transportation, and the Department of Education among others. The assistance provided by these agencies is often requested, and then reimbursed by FEMA (this is referred to as a Mission Assignment). However, in some circumstances the agencies have the authority to fund their assistance efforts through their respective budgets— even when an incident is declared under the Stafford Act. As will be shown later in this report, the assistance provided by these agencies has a bearing on how the spending caps are determined under the BCA.

The Disaster Relief Fund

As mentioned previously, the majority of disaster assistance is provided by FEMA. Once the declaration has been issued by the President, FEMA provides various forms of disaster relief through its Disaster Relief Fund (DRF). The DRF is a no-year account[4] that is used to fund response activities and pay for ongoing recovery programs. The DRF is also used to reimburse Mission Assignments to other federal agencies, and pay for declared emergencies, fire management assistance grants, and major disasters that might occur.[5]

Current budgetary practice generally consists of funding the DRF through regular appropriations acts and begins with the Administration's formulation of the budget request for the account. Among the data points used to determine the budget request are: (1) funding levels currently available in the DRF; (2) the five-year rolling average of "normal" disaster costs;[6] (3) pending recovery costs; and (4) the estimated monthly recoveries of unobligated funds.[7]

Based on these data points, since FY1998 the Administration's request for the DRF has been $1.3 billion or more each year. The average budget request between FY2000 and FY2011 is roughly $2 billion. Yet, the average the current spend-out rate[8] for the DRF has been $350 million dollars per month, which amounts to $4.2 billion a year.[9] All things being equal, at this rate the DRF may run out of funding in any fiscal year for which it is budgeted for less than $4 billion (see Figure 2).

Supplemental Appropriations

Congress provides additional budget authority to the DRF when the balance is deemed insufficient to provide for assistance and recovery projects.[10] This is primarily done through supplemental appropriations acts.[11] Disasters costing more than $500 million are considered outliers when FEMA budgets for the new fiscal year; however, this figure has been used for over a decade without being adjusted for inflation. In addition, based on data provided by FEMA, since 1992 there have been 20 declared disasters that have cost $500 million dollars or more (see *Figure 1*). These declarations are presumably considered outliers for the purposes of budgeting for disasters. However, some might argue these incidents occur too frequently to be considered as outliers and ought to be included in calculating necessary budget levels for current and prospective disaster costs.

Moreover, additional budget authority has been needed when a series of events occurs over a limited period of time. For example, in FY2008, additional budget authority for disaster assistance was needed for wildfires, floods, and hurricanes. In that year, while two hurricanes— Gustav and Ike— each exceeded the $500 million threshold, other disaster events, along with ongoing disaster recovery needs from previous events, compounded the demand for federal funding. Consequently, there has been an increased reliance on the DRF to support communities and individuals in need. This can be seen in the funding history for the DRF from FY1991 to FY2011, shown in *Figure 2*.

The prevailing trend in recent years is for the DRF to need more funding than is provided in its base budget and in the regular appropriation acts—a need generally met through supplemental appropriations legislation. The use of supplemental appropriations is of concern because traditionally they have not been subject to budgetary constraints as they frequently have been designated as emergency appropriations—allowing them to be provided in excess of spending limits otherwise established by budget legislation intended to control the size of the deficit. Some critics of past policies have asserted that

Administrations have failed to request adequate funding for the DRF in order to mask potential disaster costs and project smaller deficits in their budget documents.[12]

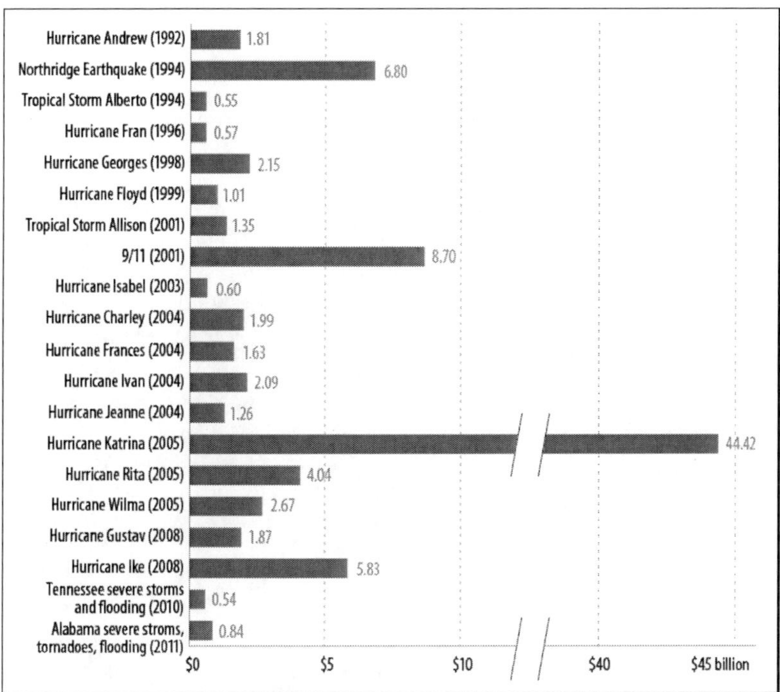

Source: Based on data provided to the authors by FEMA's Legislative Affairs Division.

Note: Spending on recovery continues for many of these events; these amounts represent a snapshot of disaster costs as of December, 2011.

Figure 1. Disasters Costing $500 Million Dollars or More,1992-2011.

In *Figure 2*, FY1998 to FY2001 seem to be outliers, in that adequate funding was requested to meet disaster needs, but when those years are more closely examined, the Administration's request for regular appropriations is actually less than $500 million in each of these years, supplemented by an emergency appropriation requested as part of the base budget—essentially, the budget process began with a built-in supplemental for disaster relief that would score outside the discretionary spending caps. This practice continued until the FY2003 request. As a result of the concern over the size of the deficit and rising level of national debt, Congress has implemented measures to limit federal spending. The BCA includes measures that limit discretionary spending while providing a mechanism that recognizes the unexpected nature of disasters and the periodic need for disaster relief funding beyond what the budget envisions.

Budget Control Act (BCA)

On August 2, 2011, the President signed into law the Budget Control Act of 2011 (P.L. 112-25), which allowed the national debt ceiling to be raised while also implementing a range of budget-controlling mechanisms.[13] As part of the legislation, caps were placed on discretionary spending for the next ten years, beginning with FY2012. If these caps are exceeded, an automatic cancellation of budget resources—known as sequestration—takes place across most discretionary budget accounts to reduce spending down to the cap.

The BCA allows for adjustments to the cap in a handful of situations, essentially raising it to allow for certain categories of spending. One of those adjustments is for emergencies, which is familiar to many observers of the budget process, but a new category of spending was defined in law for "disaster relief," allowing it to be treated separately from other emergencies.

In the past, "emergency appropriations" were often synonymous with "disaster relief."

As noted earlier, although a base level of funding was provided in regular appropriations bills for the DRF and other programs that support disaster response and recovery efforts, these accounts were often bolstered by supplemental appropriations bills as needed. Often the disaster funding was designated as an emergency appropriation, which meant that it would not count against statutory or congressional discretionary spending limits and thus did not have to be offset.Under the BCA, the discretionary spending limit can be adjusted upward to make room for an uncapped amount of emergency spending and adds the following definitions to existing budget law:

> (20) The term "emergency" means a situation that—
>> A. requires new budget authority and outlays (or new budget authority and the outlays flowing there from) for the prevention or mitigation of, or response to, loss of life or property, or a threat to national security; and
>> B. is unanticipated.

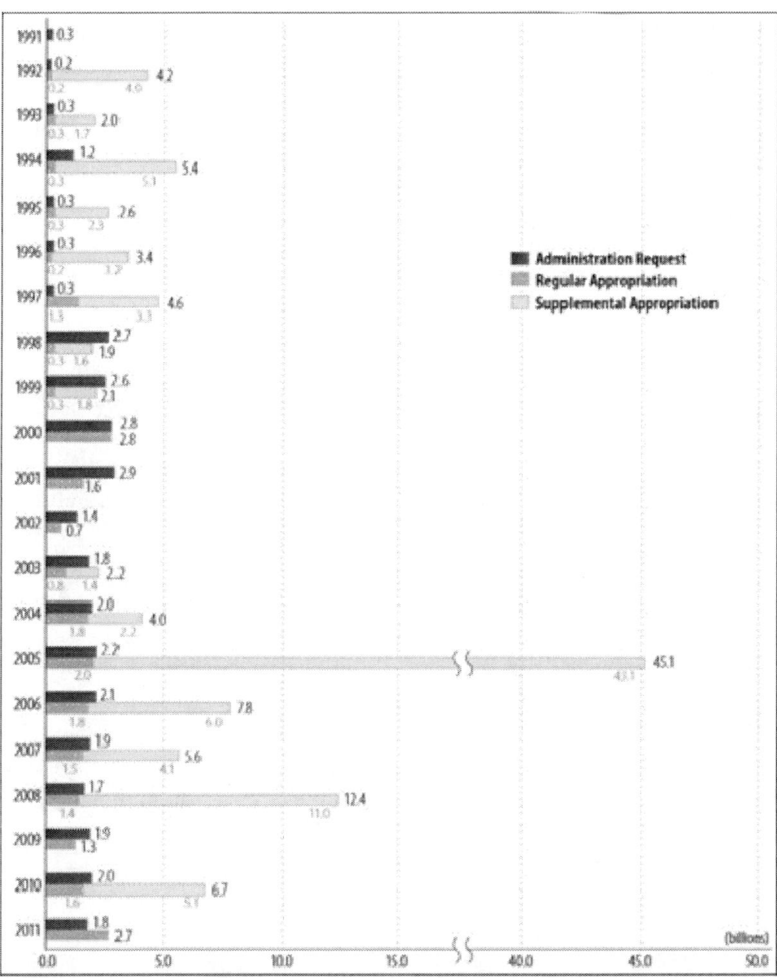

Source: CRS data using Administration budget documents and appropriations statutes.
Note: Figures have been rounded. Amounts reflect requests and appropriations in a given fiscal year, without regard to emergency designations under budget control legislation or linkage to particular disasters. Administration requests from FY1991 to FY1997 represent the previous average spending amount of $320 million during the disaster-quiescent 1980s, with the exception of 1994 when the request level reflected the estimated spending for the Midwest flooding of 1993. Based on the increased disaster activity of the1990s and outlier events such as the Mississippi flooding, Hurricane Andrew, and the Northridge earthquake, requested amounts were recalibrated for FY1998 to better address the growth in disaster spending on an annual basis. Rates of spending on disaster recovery by FEMA differ over time depending on available funding and emergency needs.

Figure 2. Disaster Relief Fund, Administration Request and Appropriation (Over FY1991-FY2011, budget authority in billions).

(21) The term "unanticipated" means that the underlying situation is—
   A. sudden, which means quickly coming into being or not building up over time;
   B. urgent, which means a pressing and compelling need requiring immediate action;
   C. unforeseen, which means not predicted or anticipated as an emerging need; and
   D. temporary, which means not of a permanent duration.[14]

Prior to enactment of the BCA, supplemental appropriations for disaster relief were often designated as emergency spending. The enactment of the BCA distinguishes disaster relief spending, though often unanticipated, as separate from emergency spending and, as a result, limits disaster relief. "Disaster relief," under the BCA, is defined as activities carried out pursuant to a determination under Section 102(2) of the Stafford Act, which authorizes the President to make a major disaster declaration.[15] The BCA, however, created a separate limited cap adjustment specifically for disaster relief, and says that funding designated as disaster relief is not eligible for the unlimited cap adjustment for emergency spending.[16]

The limit established by the BCA on adjustments to the cap for disaster relief is based on the average funding provided for disaster relief over the last ten years, excluding the highest and lowest annual amounts, calculated by the Office of Management and Budget.[17] If Congress spends less than that average on disaster relief in a given fiscal year, the cap can be further adjusted upward by the unspent amount in the following year. It is important to note that this adjustment limitation is not a restriction on disaster assistance per se—rather it is a restriction on how much the cap can be adjusted upward to accommodate the assistance. Also, spending within the cap does not require offsets. As mentioned earlier, Congress has in the past funded the DRF and other elements of the budget that provide disaster assistance with discretionary budget authority that falls within the cap.

OMB Report

The Office of Management and Budget (OMB) manages the sequestration process and the limits on adjustments available to raise the spending cap. The BCA requires OMB to annually calculate the adjusted 10-year rolling average of disaster relief spending that sets the allowable cap adjustment for disaster relief. OMB's methodology for calculating the allowable adjustment is tied to the language of the BCA's definition of "disaster relief" which includes only amounts that were appropriated or authorized through legislation that specifically referenced Section 102(2) of the Stafford Act. In its report, OMB illustrated this by comparing two similar education programs targeting students in hurricane-affected areas.[18] One program had appropriations language specifically referencing the major disaster declaration—which was counted, and one program that had language only mentioning the hurricanes rather than the disaster declaration—which was not counted. In making its calculation, OMB included funding provided through both annual and supplemental appropriations bills for 29 individual accounts managed by 11 agencies and departments. OMB has not made any other estimates of "disaster relief" spending other than those called for by the BCA.

In accordance with the BCA, OMB has calculated the allowable adjustment for FY2012 as $11.3 billion.[19] OMB will make a similar calculation each year, taking into account the latest information available on disaster funding for the 10 previous fiscal years, and excluding the highest and lowest years. As the average rolls forward, the available adjustment will change.

It is worth noting, however, that the cap is calculated in nominal dollars and does not adjust for inflation. This will become more significant over time if inflation rises, and if the allowable adjustment begins to decrease as projected in 2016.

Analysis and Potential Policy Implications

A number of policy questions are emerging as a result of the implementation of the BCA. These include what the implications are of possible failures to capture appropriate disaster relief expenditures in OMB calculations; debate over Stafford Act assistance that is excluded from the disaster relief calculations; whether severe disaster years will strain the adjustment mechanism; debate over the use of offsets; and the implications of the rising number of emergency declarations under the Stafford Act.

Expenditures Omitted from the 10-Year Average

As previously mentioned, under the BCA, future spending caps on disaster relief and OMB's methodology for calculating the allowable adjustment are based on Section 102(2) of the Stafford Act. As a result of OMB's interpretation of the definition, when OMB reviewed appropriations for inclusion in the "disaster relief" calculation, if the Stafford Act was not explicitly cited those amounts were omitted—even when the funding was clearly for response to incidents declared as major disasters (see OMB quotations below). In some cases the legislative text included "pursuant to the Stafford Act." In other cases this specific language was omitted. It is not likely that precision in the language contemplated that the wording would one day be the basis of a cap on disaster spending.

Its review resulted in this construction: when the legislative text stated the funding was pursuant to the Stafford Act, OMB included that amount in the 10-year average. On the other hand, when the legislative text made no reference to the Stafford Act—whether it referred to the declared incident or not—OMB did not include that amount in the 10-year average. OMB illustrated such omissions in the Report's methodological description. According to OMB:

> ... in determining the amount that was "provided for disaster relief" in fiscal year 2005, OMB included in the calculation the funding that the Congress appropriated ... to the Department of Education "Hurricane Education Recovery" account for "assisting in meeting the educational needs of individuals affected by hurricanes in the Gulf of Mexico" because the appropriations language specified that it was "for students attending institutions ... located in an area in which a major disaster has been declared in accordance with section 401 of the Robert T. Stafford Disaster Relief and Emergency Assistance Act."[20]

The OMB Report further states:

> OMB did not include in its calculations those amounts ... Congress appropriated in response to a presidentially-declared major disaster when such amounts were not specifically designated in statute to carry out activities pursuant to the Stafford Act and the Act itself was not specifically referenced. For example, OMB did not include in its calculations for fiscal year 2009 the appropriations ... Congress provided in December 2009 to the Department of Education "Innovation and Improvement" account "for competitive awards to local educational agencies located in counties in Louisiana, Mississippi, and Texas that were designated by ... [FEMA] as counties eligible for individual assistance due to damage caused by Hurricanes Katrina, Ike, or Gustav" because the amounts were not specified as being for activities undertaken pursuant to a major disaster declaration under the Stafford Act and the [Stafford] Act was not specifically referenced.[21]

OMB took this position despite the fact that it has not always been the practice to include a specific reference to the Stafford Act in supplemental appropriations for assistance in response to major disasters. An example of past practice is presented below.

Pre-BCA Disaster Assistance Spending

In the Disaster Relief and Recovery Supplemental Appropriations Act, 2008,[22] Title I, Chapter 1 outlines relief funds provided through the Department of Agriculture, including the following provisions (emphasis added):

> NATURAL RESOURCES CONSERVATION SERVICE
> EMERGENCY WATERSHED PROGRAM
> For an additional amount for the "Emergency Watershed Protection Program", $100,000,000, to remain available until expended, for disaster recovery operations.
> FARM SERVICE AGENCY
> EMERGENCY CONSERVATION PROGRAM
> For an additional amount for "Emergency Conservation Program", $115,000,000, to remain available until expended.
> RURAL DEVELOPMENT PROGRAMS
> RURAL DEVELOPMENT DISASTER ASSISTANCE FUND
> For grants, and for the cost of direct and guaranteed loans, for authorized activities of agencies of the Rural Development Mission Area, $150,000,000, to remain available until expended, which shall be allocated as follows: $59,000,000 for single and multifamily housing activities; $40,000,000 for community facilities activities; $26,000,000 for utilities activities; and $25,000,000 for business activities: Provided, That such funds shall be for areas affected by hurricanes, floods, and other natural disasters occurring during 2008 for which the President declared a major disaster under title IV of the Robert T. Stafford Disaster Relief and Emergency Assistance Act of 1974: Provided further, That the cost of such direct and guaranteed loans, including the cost of modifying loans, shall be as defined in section 502 of the Congressional Budget Act of 1974: Provided further, That the Secretary of Agriculture may reallocate funds made available in this paragraph among the 4 specified activities, if the Secretary notifies the Committees on Appropriations of the House of Representatives and the Senate not less than 15 days prior to such reallocation.
> In addition, for an additional amount for grants, and for the cost of direct and guaranteed loans, for authorized activities of the Rural Housing Service, $38,000,000, to remain available until expended, for single and multi-family housing activities: Provided, That such funds shall be for

areas affected by Hurricanes Katrina and Rita: Provided further, That the cost of such direct and guaranteed loans, including the cost of modifying loans, shall be as defined in section 502 of the Congressional Budget Act of 1974.

Of all the appropriations listed, only the provisions in bold would be counted by OMB for purposes of calculating the cap on the adjustment for disaster relief as defined under the Budget Control Act. Only the Rural Development Disaster Assistance Fund appropriation specifically noting the declaration of a major disaster under the Stafford Act meets the standard described in OMB's report. The other provisions, mentioning the storms that were the root cause of the declaration, or the intent that the funds be for "disaster recovery" would likely not be adequate to meet the OMB methodology for accounting for disaster relief spending.

Disaster Relief Spending Under the BCA

In the "minibus" legislation, P.L. 112-55, provisions providing disaster relief under some of these same accounts were written as follows (emphasis added):

Section 735. There is hereby appropriated for the 'Emergency Conservation Program', for necessary expenses resulting from a major disaster declared pursuant to the Robert T. Stafford Disaster Relief and Emergency Assistance Act (42 U.S.C. 5121 et seq.), $122,700,000, to remain available until expended: Provided, That the preceding amount is designated by the Congress as being for disaster relief pursuant to section 251(b)(2)(D) of the Balanced Budget and Emergency Deficit Control Act of 1985: Provided further, That there is hereby appropriated for the 'Emergency Forest Restoration Program', for necessary expenses resulting from a major disaster declared pursuant to the Robert T. Stafford Disaster Relief and Emergency Assistance Act (42 U.S.C. 5121 et seq.), $28,400,000, to remain available until expended: Provided further, That the preceding amount is designated by the Congress as being for disaster relief pursuant to section 251(b)(2)(D) of the Balanced Budget and Emergency Deficit Control Act of 1985: Provided further, That there is hereby appropriated for the 'Emergency Watershed Protection Program', for necessary expenses resulting from a major disaster declared pursuant to the Robert T. Stafford Disaster Relief and Emergency Assistance Act (42 U.S.C. 5121 et seq.), $215,900,000, to remain available until expended: *Provided further, That the preceding amount is designated by the Congress as being for disaster relief pursuant to section 251(b)(2)(D) of the Balanced Budget and Emergency Deficit Control Act of 1985.*

All of this funding would be considered by OMB as disaster relief due to the citations of major disasters under the Stafford Act, as well as the specific proviso in bold declaring Congressional intent that it be categorized as such.

It could be argued that a more precise 10-year average of disaster assistance would include all spending for major disasters regardless of whether the legislative text referred to the Stafford Act.

There are at least two changes that could be made that would help ensure a more accurate calculation of the 10-year average spent on disaster relief for use as a cap adjustment under the BCA. First, Congress could change their practices and provide the Stafford Act designation in all future appropriations legislation. However, it is important to note that at the time of appropriation, it is not always clear if funding is going to a major disaster, thus meriting a designation. Second, Congress could amend the BCA to require that OMB recalculate "disaster relief" amounts based on a broader methodology. Both of these changes would likely result in a higher, and arguably more accurate yearly total of disaster relief, and a larger allowable adjustment for disaster relief under the BCA than under current practices.

For example, Hurricane Katrina was declared a major disaster on August 29, 2005. Since then, Congress has provided disaster assistance through several supplemental appropriations and numerous federal programs. Large incidents like Katrina often receive assistance from the federal government years after the incident—the appropriations impact the budget for disaster assistance as large infrastructure and mitigation projects are completed and reimbursed, yet because this funding was appropriated without direct reference to a Stafford Act declaration, it is not factored into the calculation for disaster relief.

Other Types of Excluded Stafford Assistance

The BCA excludes other types of assistance provided under the Stafford Act. These are emergency declarations provided under Section 102(1) and Fire Management Assistance Grants provided under Section 420(a). Emergency declarations authorize activities that can help states and communities carry out essential services during emergency situations.[23] Emergencies can also be declared prior to an incident, at the request of the governor, to save lives and prevent loss.[24] For example, emergency declarations have been declared prior to a hurricane making landfall to help state and local governments take necessary measures (evacuation assistance, placement of response resources, etc.).[25] Unlike major disasters, the President does have the authority to declare an emergency without a governor's request when the incident involves a subject area where the "Federal government exercises exclusive or primary responsibility and authority."[26]

Compared to major disaster declarations, emergency declarations are generally considered a minor expense (congressional notification is required when spending for an emergency exceeds $5 million); however, numerous declarations can be declared in a year and, like major disasters, they are funded through the DRF. In 2005, 68 emergency declarations were declared, 50 of which were for each individual state to help relocate Hurricane Katrina victims who were displaced by the storm. In addition, since Hurricane Katrina, the federal government has increased its efforts to pre-position resources before a hurricane makes landfall. If this trend continues the cost associated with emergency declarations may increase due to the more comprehensive preparations.

In the OMB report, the spending levels on disaster relief from the DRF in OMB's accounting is less than the total amount expended from the DRF in the years reported. This difference may include the omission of expenditures for emergency declarations and declarations for the Fire Management Assistance Grant Program.[27]

Declarations for the Fire Management Assistance Grant Program include equipment, personnel, and supplies to states and localities for the mitigation, management, and control of fires that threaten to become a major disaster.[28] As with emergency declarations, declarations for the Fire Management Assistance Grant Program are relatively modest in cost when compared to major disaster declarations. A review of declarations under the Fire Management Assistance Grant Program shows the most expensive year was 1998, in which 53 declarations were made, accounting for obligations of roughly $105 million.[29]

Because emergency and fire declarations derive funding from the DRF, it could be argued that excluding them from the ten-year average calculation for disaster relief generates an artificially low result. It could be further argued that including emergency and fire declarations would more accurately forecast federal disaster expenses. Although it is likely that including all federal assistance for emergency and disaster relief would increase the ten-year average, the size of the increase would depend on the new methodology used to calculate the amount of assistance provided.

Funding for Severe Disaster Years

Congress provided additional budget authority for disaster assistance in 10 appropriation laws following the hurricanes of 2005. While OMB removed the $37 billion spent on disaster relief in 2005 as an outlier when calculating the allowable adjustment for disaster relief to the cap, response and recovery to these storms went well beyond that first year. Appropriations for recovery from these storms between FY2006 and FY2010 was still substantial—$32 billion was spent in FY2006, in great part because of those ongoing recovery efforts.

The sizeable initial disaster relief expenditures for Hurricane Katrina and the other 2005 storms will begin to lose relevance in calculating the allowable adjustment for disaster assistance for FY2016, and will no longer impact calculations for the allowable adjustment in FY2017. Once FY2005 and FY2006 rotate out, there will be a corresponding drop in the allowable disaster assistance adjustments. The reduction in the allowable adjustment will be more significant if disaster spending is below the 10-year average in the intervening years. In a scenario where disaster spending stays at the 10-year average level, the allowable adjustment will fall by $2.2 billion from FY2015 to FY2016, and then by another $2.9 billion from FY2016 to FY2017—a reduction of 41% in just two years. Moreover, as the Administration and Congress work to spend under the adjustment, the cap could continue to decrease, increasing the likelihood of a bad storm year significantly straining the budget mechanisms in place at that time.

Some policymakers might welcome such a series of developments, arguing the purpose of the BCA is to rein in deficit spending by either keeping spending under the caps, or by triggering a sequestration. Others might contend that the limitations on disaster spending are too severe, given the unpredictable demand for disaster relief in the context of a very tight budget where offsets may be difficult to come by without significant impacts on government operations. Congress would have to spend well above the eligible adjustment in at least two years before FY2016 to maintain a consistent funding baseline for disaster assistance—decisions that could constrain the rest of the discretionary budget.

It is possible that the Administration and Congress could rely on the available adjustment to fund all accounts where disaster relief is appropriated, rather than relying on discretionary spending under the cap to do so. This may be a particularly attractive option for appropriators as the discretionary budget becomes more constrained. If this pattern of funding becomes the norm, and a severe disaster year demands increased disaster relief, as Congress seeks more funds, there are at least four possible outcomes: (1) discretionary spending cuts, either by Congress or through sequestration, to "pay for" the additional assistance, (2) designation of disaster assistance as emergency funding, despite the definitions in the BCA, (3) creation of designated revenue raisers by the Congress to finance extra disaster relief spending, or (4) renegotiation of the underlying budget control laws.

Another potential issue is the impact of a mega-disaster might have on the federal budget. For example, the FY2005 supplemental for the 2005 hurricane season was $45.1 billion. If an event with similar or even greater damage costs, such as a New Madrid earthquake, occurred under the current budget constraints set forth in the BCA, Congress could find itself in need of roughly hundreds of billions in offsets to pay for disaster relief or possibly facing a sequestration of similar size if it adheres to the BCA's definitions.[30] One might question whether the federal budget could absorb a sequestration of this magnitude without inflicting a severe impact on a national economy already shocked by the direct impact of the disaster. An alternative to that possibility is that, if such conditions arise, Congress may choose to reach an agreement outside of the timelines and constraints currently set forth in the BCA.

Contention over Disaster Assistance Offsets

In the fall of 2011, there was extensive public debate over the possible requirement of offsets for disaster assistance. Those opposed to the use of offsets argue that their use could politicize disaster assistance by allowing policymakers to target certain programs for the needed spending reduction. Assistance to disaster victims could be delayed while Congress debates the issue. Opponents have also argued that emergency funding for other endeavors, such as war funding, have not faced the same requirement.

Those in favor of offsetting disaster assistance argue that offsets do not deny disaster victims aid; they merely provide a way of doing so without increasing the deficit. Proponents also argue that the concern over delayed disaster assistance is without merit. As demonstrated in *Table 1*, while it may take some time to provide relatively smaller incidents such as the Nisqually Earthquake with supplemental funding, Congress has responded to the needs of disaster victims by appropriating additional funds for disaster relief in a matter of days as with the September 11th terrorist attacks and Hurricane Katrina. It should be noted however, the actions represented in *Table 1* were emergency spending actions that were not subject to offsets. Whether congressional action would be as rapid under the BCA framework is uncertain.

Table 1. Supplemental Funding for Large-Scale Disasters

Event	Date of Declaration	Date of Enactment	Days
Hurricane Katrina	August 29, 2005	September 2, 2005	3
Hurricane Isabel	September 18, 2003	September 30, 2003	12
9/11 Terrorist Attacks	September 11, 2001	September 18, 2001	7
Nisqually Earthquake	March 1, 2001	July 24, 2001	114
Hurricane Floyd	September 16, 1999	October 20, 1999	34
Northridge Earthquake	January 17, 1994	February 12, 1994	26
Midwest Floods	June 11, 1993	August 12, 1993	62
Hurricane Andrew	August 23, 1992	September 23, 1992	31
Hurricane Hugo	September 20, 1989	September, 29, 1989	9

Source: CRS Report R40708, Disaster Relief Funding and Emergency Supplemental Appropriations, by Bruce R. Lindsay and Justin Murray.

Note: Table 1 reflects the number of days it took to enact the first supplemental appropriation after the incident was declared a major disaster. Some incidents (such as Hurricane Katrina) received more than one supplemental appropriation for disaster relief.

Sequestration may seem more appealing to some Members rather than finding offsets for disaster assistance if the allowable adjustment is inadequate and Congress chooses to not use emergency appropriations to support the DRF. Although the net accounting effect is the same over the medium term, sequestration involves automatic, largely across-the-board spending reductions after the fact, rather than a specific Congressional decision, affirmed by votes of the House and Senate and signed by the President to reduce funding for a specific program or programs with allies and stakeholders who may be provoked to action. The potential risks incurred by an automatic across-the-board cancellation of budget authority across the government regardless of the possible effect on national priorities should not be discounted, however.

Increasing Declarations

Since FEMA's first full year of operations (1979) there has been a steady increase in the number of emergency and major disaster declarations. It is unclear what is causing the increase. On the one hand, it could be the result of more incidents. On the other hand, it could be the result of an increase in incidents for which a request for assistance is made (in other words, there is no increase in the number of incidents, rather, there is an increase in requests for federal assistance). The result could also be caused by a combination of the two, as well as by some other undetermined cause. However, while the number of declarations is often a focus for criticism, it is the costs within the declared events (determinations on eligible disaster spending) that can drive the higher disaster spending amounts.

The BCA provides a mechanism designed to reduce the impact of disaster relief spending on the national debt, but does not provide a means for limiting or reducing federal expenditures on disaster assistance. If declarations continue to increase unabated, the federal budget may have to absorb more and more of the costs associated with disaster assistance. Some might argue that in addition to spending adjustments for disaster assistance, other policies designed to reduce federal expenditures for assistance should also be pursued. These policy options include strengthening declaration criteria, reducing the federal cost-share for incidents, and creating incentives that would encourage states to pursue more robust preparedness and mitigation measures.[31]

CONCLUDING OBSERVATIONS

The debate over the use of mechanisms to reduce the impact of disaster assistance on the deficit, such as the use of offsets, was temporarily moved to the back burner when FEMA deobligated funds designated for projects that came in under budget, providing, in conjunction with delays in funding for pending recovery projects to slow the use of resources, enough resources to fund the DRF to last until the beginning of the 2012 fiscal year.

The debate over disaster assistance may take place again in the near future due to a number of factors. While the current continuing resolution (P.L. 112-36) provided $2.65 billion in FY2012 for disaster relief, as previously mentioned, the average spend-out rate for the DRF is $350 million per month—or $4.2 billion a year. If the average spending remains the same there may be another shortfall in the DRF some time before the end of FY2012. Moreover, in addition to recovery costs associated with Hurricane Katrina, significant costs were incurred in FY2011 that will also draw funding from the DRF in FY2012. These include Hurricane Irene (estimated at roughly $1.5 billion), the 2011 fires, the Mid-Atlantic Earthquake of 2011, and Tropical Storm Lee. As mentioned previously, the BCA is designed to allow for a limited amount of additional spending on disaster relief beyond the discretionary spending limits it sets out. If the costs of disaster assistance continue to grow, other parts of the federal budget will need to absorb those costs if they exceed the adjustments. Policymakers may need to consider additional cost-saving measures to prevent this from occurring.

While the BCA may help curb or contain the impact of traditional disaster spending, the implications of a truly catastrophic incident such as Hurricane Katrina are unclear. If such conditions arise, Congress may choose to reach an agreement outside of the timelines and constraints currently set forth in the BCA.

End Notes

[1] Data and information on disasters can be found at, NOAA: National Climate Data Center, National Climate Data Center Billion Dollar U.S. Weather/Disasters Page, NCDC Billion Dollar U.S. Weather/Climate Disasters page, June 17, 2011, http://www.ncdc.noaa.gov/oa/reports/ billionz.html.

[2] 42 U.S.C. 5121 et seq. For further analysis on Stafford Act disaster assistance see CRS Report RL33053, Federal Stafford Act Disaster Assistance: Presidential Declarations, Eligible Activities, and Funding, by Francis X. McCarthy.

[3] For further analysis on emergency and disaster declarations see CRS Report RL34146, FEMA's Disaster Declaration Process: A Primer, by Francis X. McCarthy.

[4] While most appropriations expire after a set period of time, no-year appropriations are available until expended. This is helpful in disaster recovery since infrastructure repair and mitigation projects can stretch out over several years.

[5] Fire Management Assistance Grants (FMAGs) and emergencies under the Stafford Act are discussed later in the report.

[6] Normal disasters are declared incidents that cost less than $500 million dollars. Disasters costing over $500 million are considered outliers and are removed from the calculation.

[7] For example, when a recovery project is completed for less than the estimated cost.

[8] The spend-out rate refers to the amount of money paid out of the account for a given period of time.

[9] Based on a CRS discussion with FEMA staff from the Office of the Chief Financial Officer.

[10] Congress also appropriates disaster funds to other accounts administered by other federal agencies pursuant to federal statutes that authorize specific types of disaster relief.

[11] For further analysis on emergency supplemental appropriations see CRS Report R40708, Disaster Relief Funding and Emergency Supplemental Appropriations, by Bruce R. Lindsay and Justin Murray.

[12] For example, see Office of Management and Budget, A New Era of Responsibility: Renewing America's Promise, Washington, DC, February 26, 2009, p. 36, http://www.gpoaccess.gov/usbudget/fy10/pdf/fy10-newera.pdf.

[13] For further analysis of the Budget Control Act see CRS Report R41965, The Budget Control Act of 2011, by Bill Heniff Jr., Elizabeth Rybicki, and Shannon M. Mahan.

[14] Budget Control Act of 2011, P.L. 112-25, §102.

[15] Under a major disaster declaration, state and local governments and certain nonprofit organizations are eligible (if so designated) for assistance for the repair or restoration of public infrastructure such as roads and buildings. A major disaster declaration may also include temporary housing, unemployment assistance, crisis counseling for families and individuals, and community disaster loans for local governments. The governor of the

impacted state requests the types of assistance considered necessary to address the needs of the state.

[16] Budget Control Act of 2011, P.L. 112-25, §101.

[17] Ibid.

[18] Office of Management and Budget, OMB Report on Disaster Relief Funding to the Committees on Appropriations and the Budget of the U.S. House of Representatives and the Senate, Washington, DC, September 1, 2011, pp. 2-3.

[19] Office of Management and Budget, OMB Report on Disaster Relief Funding to the Committees on Appropriations and the Budget of the U.S. House of Representatives and the Senate, Washington, DC, September 1, 2011, p. 1.

[20] Office of Management and Budget, OMB Report on Disaster Relief Funding to the Committees on Appropriations and the Budget of the U.S. House of Representatives and the Senate, September 1, 2011, p. 2, http://www.whitehouse.gov/sites/default/files/omb/assets/legislative_reports/disaster_relief_report_sept2011.pdf.

[21] Ibid., p. 2.

[22] Division B of P.L. 110-329, 122 Stat. 3585 et seq.

[23] For example, food, sheltering and medical care.

[24] For example, evacuations and setting up shelters.

[25] Recent examples of pre-event declarations include emergency declarations prior to Hurricanes Katrina, Rita, and Gustav making landfall (emergency declarations 3212, 3260, and 3290 respectively).

[26] 44 CFR 206.35(d). This category would likely include acts of terrorism.

[27] OMB did not respond to CRS inquiries about the details of its methodology.

[28] 42 U.S.C. 5187.

[29] DHS/FEMA, Calendar Year Summary of Obligations, 1988-2010.

[30] W. Barksdale Maynard, "When the Big Muddy Ran Backward," The Washington Post, January 31, 2012, p. E1.

[31] For further analysis on emergency and major disaster declarations see CRS Report RL34146, FEMA's Disaster Declaration Process: A Primer, by Francis X. McCarthy. For further analysis on FEMA cost-shares see CRS Report R41101, FEMA Disaster Cost-Shares: Evolution and Analysis, by Francis X. McCarthy.

In: The Community Disaster Loan Program   ISBN: 978-1-62417-643-2
Editor: Felix P. Cardano   © 2013 Nova Science Publishers, Inc.

Chapter 3

OMB REPORT ON DISASTER RELIEF FUNDING TO THE COMMITTEES ON APPROPRIATIONS AND THE BUDGET OF THE U.S. HOUSE OF REPRESENTATIVES AND THE SENATE[*]

The Office of Management and Budget

The Honorable Harold Rogers
Chairman
Committee on Appropriations
U.S. House of Representatives
Washington, DC 20515

Dear Mr. Chairman:

Enclosed please find the *OMB Report on Disaster Relief Funding to the Committees on Appropriations and the Budget of the U.S. House of Representatives and the Senate.*

It has been prepared pursuant to section 251 of the Balanced Budget and Emergency Deficit Control Act of 1985 (BBEDCA), as amended. As you know, when the Congress passed the Budget Control Act (BCA) last month, it

---

[*] This is an edited, reformatted and augmented version of The Office of Management and Budget publication, dated September 1, 2011.

included a provision to account for disaster relief spending. For purposes of fiscal year 2012, Congress allowed for the discretionary cap total to be raised by no more than the average funding provided for disaster relief over the previous 10 years, excluding the highest and lowest years. Consistent with the requirements of the BCA, this report sets forth that ceiling.

Also, as referenced in the *OMB Sequestration Update Report to the President and Congress for Fiscal Year 2012,* this report includes a discussion of the preview estimate of the disaster funding adjustment for fiscal year 2012 as required by Section 254(e) of BBEDCA, as amended. The average funding provided for disaster relief over the previous 10 years (excluding the highest and lowest years) is $11.5 billion for fiscal year 2011, and $11.3 billion for fiscal year 2012.

The latter amount represents a ceiling on the potential disaster adjustment for fiscal year 2012, and the actual level of the adjustment will be decided through the fiscal year 2012 appropriations process. The Administration assessed what is required to properly fund the Disaster Relief Fund and known disaster relief needs as identified under existing law including enduring costs from previous disasters such as Hurricane Katrina which hit in 2005 and the tornadoes that struck Missouri earlier this year.

This analysis supports a disaster relief adjustment of $5.2 billion for fiscal year 2012. This amount does not account for the devastation Hurricane Irene has inflicted on communities up and down the eastern part of the country. At this writing, rescue efforts are still underway and floodwaters are, in some areas, still rising. Funds are being delivered in real time to meet pressing needs. With four major disasters already declared, we know that the costs of this storm will be significant.

We are working closely with the Federal Emergency Management Agency and the Department of Homeland Security to determine if additional resources for 2011 will be necessary.

There is no question, however, that additional funds will be required to assist the thousands of Americans affected by Hurricane Irene, on top of the $5.2 billion identified under current law to properly fund known disaster needs for fiscal year 2012. As we gather data about the extent of damage and need caused by Hurricane Irene, the Administration promptly will determine that amount and provide it as soon as possible to the Congress during its appropriations process.

Providing disaster relief to our fellow Americans is a commitment that crosses all boundaries of geography and political party.

The Administration looks forward to working with the Congress to continue this long tradition of providing for our neighbors as they recover and rebuild their homes and communities.

Sincerely,

Jacob J. Lew Director

I. INTRODUCTION

P.L. 112–25, the Budget Control Act of 2011 (BCA), amended the Balanced Budget and Emergency Deficit Control Act of 1985 (BBEDCA), by reinstating the discretionary spending limits that had expired after 2002 and by creating ad-justments to those limits. Section 251(b)(2)(D)(ii) of BBEDCA requires the Office of Management and Budget (OMB) to submit a report to the Committees on Appropriations and the Budget in the U.S. House of Representatives and the Senate within 30 days of the enactment of the BCA on the average funding provided for disaster relief over the previous 10 years (excluding the highest and lowest years). In addition, section 254(e) of BBEDCA, as amended by section 103(2) of the BCA, requires OMB to include in its Sequestration Update Report a "preview estimate of the adjustment for disaster funding for the upcoming fiscal year," in this case FY 2012.

The average funding provided for disaster relief over the previous 10 years (excluding the highest and lowest years) is $11.5 billion for fiscal year 2011 and $11.3 billion for fiscal year 2012. There is an inadequate basis to estimate the final adjustment for fiscal year 2012 because of the current status of the appropriations process and the newness of the disaster relief adjustment in BBEDCA. The amount of $11.3 billion represents a ceiling on the potential disaster adjustment, but the actual level of the adjustment will be decided through the fiscal year 2012 appropriations process. Applying OMB's methodology for determining prior year levels of funding for disaster relief to the specific language in the President's fiscal year 2012 budget request and other known disaster relief needs as identified in prior law would support a disaster relief adjustment equal to $5.2 billion. In addition, there is no question that additional funds will be required to assist with the response to Hurricane Irene, on top of the $5.2 billion. As the Administration gathers data about the extent of the damage caused by Hurricane Irene, the needed amounts will promptly be determined and provided to the Congress during its appropriations process this month.

II. METHODOLOGY

Section 251(b)(2)(D)(ii) of BBEDCA requires OMB to "report to the Committees on Appropriations and Budget in each House *the average* calculated pursuant to [section 251(b)(2)(D)(i)(II)], not later than 30 days after the date of the enactment of the Budget Control Act of 2011." The "average" referenced in Section 251(b)(2)(D)(i)(II) is "the average as calculated in [section 251(b)(2)(D)(i)(I)]" for the preceding fiscal year.

Section 251(b)(2)(D)(i)(I) requires the calculation of "the average funding *pro- vided for disaster relief* over the previous 10 years, excluding the highest and lowest years." "Disaster relief" is defined in section 251(b)(2)(D)(iii) as "activities carried out pursuant to a determination under section 102(2)" of the Robert T. Stafford Disaster Relief and Emergency Assistance Act (Stafford Act). A "determination under section 102(2)" means the Presidential declaration of a "major disaster."[1]

In accordance with section 251(b)(2)(D)(ii), as applied to fiscal year 2012, this report describes, for the preceding fiscal year (FY 2011), the average funding provided for disaster relief over the previous 10 years (FY 2001–2010), excluding the highest (FY 2005) and the lowest (FY 2003) years. This report also provides the same average funding information for FY 2002 through 2011 (excluding the highest and the lowest years).

The statutory definition of "disaster relief" makes no distinction between annual and supplemental appropriations. In the absence of any such specific language, and given that a large portion of disaster relief funding is provided through supplemental appropriations, the plain reading of the statute is that it includes appropriations for disaster relief provided in any appropriations bill. Thus, OMB's calculations include both annual and supplemental appropriations.

In determining the average amount of annual funding "provided for disaster relief" over the previous 10 years (excluding the highest and lowest years), OMB included in these calculations only those amounts that the Congress, in an appropriations or authorization act in which the Stafford Act was specifically referenced, provided expressly for activities relating to a Presidential declaration of a "major disaster" under the Stafford Act. In other words, funding was included in OMB's calculations only if it was specifically appropriated for activities undertaken pursuant to a major disaster declaration under the Stafford Act (and the appropriating language referenced the Act) or was obligated pursuant to specific authority in authorizing language that provided that funding could be used for such purposes (and such language referenced the Act). 2

For example, in determining the amount that was "provided for disaster relief" in fiscal year 2005, OMB included in the calculation the funding that the Congress appropriated in December 2005 to the Department of Education "Hurricane Education Recovery" account for "assisting in meeting the educational needs of individuals affected by hurricanes in the Gulf of Mexico" because the appropriations language specified that it was "for students attending institutions of higher education...that are located in an area in which a major disaster has been declared in accordance with section 401 of the Robert T. Stafford Disaster Relief and Emergency Assistance Act."

Conversely, when determining the amounts that were "provided for disaster relief," OMB did not include in its calculations those amounts that the Congress appropriated in response to a presidentially-declared major disaster when such amounts were not specifically designated in statute to carry out activities pursuant to the Stafford Act and the Act itself was not specifically referenced. For example, OMB did not include in its calculations for fiscal year 2009 the appropriations that the Congress provided in December 2009 to the Department of Education "Innovation and Improvement" account "for competitive awards to local educational agencies located in counties in Louisiana, Mississippi, and Texas that were designated by the Federal Emergency Management Agency (FEMA) as counties eligible for individual assistance due to damage caused by Hurricanes Katrina, Ike, or Gustav" because the amounts were not specified as being for activities undertaken pursuant to a major disaster declaration under the Stafford Act and the Act was not specifically referenced.

III. DISCUSSION OF THE NUMBERS AND AVERAGE FUNDING CALCULATION

Through the application of the criteria developed using the interpretation discussed above, OMB compiled a data set that includes 29 individual accounts across 11 departments and agencies. Historical appropriated funding levels were collected for each account spanning fiscal years 2001 through

2011. Each account was broken down into two components representing base/annual appropriations and supplemental appropriations. For a full listing of the accounts and associated funding levels, see the appendix.

The calculation begins by summing the annual and supplemental accounts to the agency level.

Table 1. Summary of average funding provided for disaster relief for fy 2001 through fy 2010
(Discretionary budget authority, in millions of dollars)

Fiscal Year	Budget Authority
2001	$4,203
2002	$12,454
2003	$1,852
2004	$7,558
2005	$37,157
2006	$31,944
2007	$5,451
2008	$21,365
2009	$2,743
2010	$6,029
2011	$2,475

FY 2001–2010	
Total Budget Authority	$130,756
Low (FY 2003)	$1,852
High (FY 2005)	$37,157
Average (dropping high/low)	$11,468

For each fiscal year, respective agencies' aggregated disaster relief funding levels are added together to reach a top-line appropriated amount for disaster relief for that particular fiscal year. These nominal figures are presented in Table 1. As required by BBEDCA, an average annual amount for disaster relief is calculated over a ten-year period. BBEDCA prescribes dropping the years with the lowest and highest aggregated funding levels from the average calculation in an attempt to reduce the leveraging effect that potential outliers could have on a straightforward computation of the mean – essentially calculating an eight-year average. In Table 1, the result of the average calculation is depicted for FY 2001 through FY 2010. The total budget authority appropriated for disaster relief over this period is $130.8 billion. The low value dropped was for fiscal year 2003 ($1.9 billion), and the high value dropped was for fiscal year 2005 ($37.2 billion). *The average annual appropriated funding level for disaster relief for FY 2001 through FY 2010 (excluding the highest and lowest years) is $11.5 billion.*

Through implementation of the evaluative criteria, there are several characteristics of the data and its treatment that are worth mentioning. First, several accounts contributed no appropriated funding, indicating that, while the accounts have no historical obligations for disaster relief within the time period studied, they have the underlying statutory authority to obligate appropriated funds for this activity. Those accounts, although included in our analysis, are not listed in the appendix to this the report. They include the Department of Agriculture's Office of Inspector General account, the Department of Health and Human Services' National Institute of Environmental Health Services account, the Department of Housing and Urban Development's Home Investment Partnerships Fund account, the Department of the Interior's Operation of Indian Programs account, and the Army Corps of Engineers' Mississippi River and Tributaries account and Operation and Maintenance account.

Second, included in the data are funding for two loan accounts—FEMA's Disaster Assistance Direct Loan Program and the Small Business Administration's Disaster Loans Program. The amounts listed are a reflection of the appropriated funding levels for the accounts' subsidy and administrative expenses.

Finally, for all appropriated budget authority captured in this exercise, OMB consistently recorded pre-transfer funding levels. This is noteworthy for accounts like FEMA's Disaster Relief Fund that reimburse many agencies for their respective disaster-related activities through mission assignments. Consequently, recipient accounts were identified to prevent double counting.

IV. PREVIEW ESTIMATE OF THE DISASTER FUNDING ADJUSTMENT FOR FY 2012

As discussed in the *OMB Sequestration Update Report to the President and Congress for Fiscal Year 2012*, section 254(e) of BBEDCA, as amended by section 103(2) of the BCA, requires OMB to include in its Sequestration Update Report a "preview estimate of the adjustment for disaster funding for the upcoming fiscal year," in this case FY 2012. At the time the Sequestration Update Report was being prepared, OMB was working to complete this report, and OMB determined it would discuss the preview estimate of the disaster funding adjustment in this report.

Section 251(b)(2)(D)(i) of BBEDCA states that the adjustment shall be the total of "appropriations for discretionary accounts that the Congress designates as being for disaster relief in statute," subject to a ceiling (*i.e.*, the maximum adjustment) calculated pursuant to section 251(b)(2)(D)(i)(I) and (II). At this time, no appropriations bills for 2012 have been enacted, and the Congress has not designated any appropriations as being for disaster relief pursuant to BBEDCA; accordingly, there is no adequate basis to estimate the final adjustment.

Section 251(b)(2)(D)(i)(I) and (II) of BBEDCA provides that the ceiling for the disaster relief adjustment is calculated by adding the average funding provided for disaster relief over the previous 10 years (excluding the highest and lowest years) to an amount equal to "the difference between the enacted amount and the allowable adjustment as calculated [for the prior fiscal year]." The calculation of the ceiling in FY 2012 is outlined in Table 2. The total budget authority appropriated for disaster relief over this period is $129.0 billion. Just as in the calculation depicted in Table 1, the low value dropped was for fiscal year 2003 ($1.9 billion), and the high value dropped was for fiscal year 2005 ($37.2 billion), providing an average of $11.3 billion.

In the prior fiscal year (FY 2011) there was no allowable adjustment, as there was no spending limit pursuant to BBEDCA that could be adjusted. Therefore, the ceiling for the disaster relief adjustment in fiscal year 2012 is equal to the average funding provided for disaster relief for FY 2002 through FY 2011 (excluding the highest and lowest years), or $11.3 billion. However, no actual adjustment in that amount, or any amount, would occur unless the Congress enacts appropriations that it specifically "designates as being for disaster relief" pursuant to BBEDCA.

Applying OMB's methodology for determining prior year levels of funding for disaster relief to the specific language in the President's fiscal year 2012 budget request and other known disaster relief needs as identified in prior law would support a disaster relief adjustment equal to $5.2 billion. In addition, there is no question that additional funds will be required to assist with the response to Hurricane Irene, on top of the $5.2 billion. As the Administration gathers data about the extent of the damage caused by Hurricane Irene, the needed amounts will promptly be determined and provided to the Congress during its appropriations process this month.

Table 2. Summary of average funding provided for disaster relief for fy 2002 through fy 2011
(Discretionary budget authority, in millions of dollars)

Fiscal Year	Budget Authority
2001	$4,203
2002	$12,454
2003	$1,852
2004	$7,558
2005	$37,157
2006	$31,944
2007	$5,451
2008	$21,365
2009	$2,743
2010	$6,029
2011	$2,475

FY 2002–2011	
Total Budget Authority	$129,028
Low (FY 2003)	$1,852
High (FY 2005)	$37,157
Average (dropping high/low)	$11,252

APPENDIX

detailed table of disaster relief funding for fy 2001 through fy 2011
(Discretionary budget authority, in millions of dollars)

	2001	2002	2003	2004	2005	2006	2007	2008	2009	2010	2011	2001–2011
DEPARTMENT OF AGRICULTURE												
Farm Service Agency, Salaries and Expenses, Emergency Farm Loans (12-0600-0-1-351):												
Base												
Supplemental							0.308					0.308
Farm Service Agency, USDA Supplemental Assistance, Livestock Assistance and Dairy Disaster Assistance (12-2701-0-1-351):												
Base												
Supplemental							109					109
Rural Development, Rural Development Disaster Assistance Fund (12-0400-0-1-453):												
Base												
Supplemental								150				150
Office of the Secretary, Other Emergency Appropriations (12-9913-0-1-352):												
Base												
Supplemental							40					40
Subtotal, Department of Agriculture												
Base												
Supplemental							149	150				299

detailed table of disaster relief funding for fy 2001 through fy 2011
(Discretionary budget authority, in millions of dollars) (Continued)

	2001	2002	2003	2004	2005	2006	2007	2008	2009	2010	2011	2001–2011
DEPARTMENT OF COMMERCE												
Economic Development Administration, Economic Development Assistance Programs (13-2050-0-1-452):												
Base												
Supplemental								500		49		549
DEPARTMENT OF EDUCATION												
Office of Federal Student Aid, Student Financial Assistance (91-0200-0-1-502):												
Base						8						8
Supplemental												
Office of Postsecondary Education, Higher Education, Hurricane Aid for Postsecondary Institutions (91-0201-0-1-502):												
Base												
Supplemental							30	15				45
Hurricane Education Recovery (91-0013-0-1-500):												
Base												
Supplemental						1,885	30					1,915
Elementary and Secondary Education, School Improvement Programs, Education for Homeless Children and Youth (91-1000-0-1-501):												
Base												
Supplemental									15			15
Subtotal, Department of Education												
Base						8						8
Supplemental						1,885	60	15	15			1,975
DEPARTMENT OF HEALTH AND HUMAN SERVICES												
Administration for Children and Families, Social Services Block Grant (75-1534-0-1-505):												
Base												
Supplemental									600			600
Children and Family Services, Head Start, Disaster Human Services Case Management (75-1536-0-1-506):												
Base									2	2	2	6
Supplemental												
Administration on Aging, Aging Services Program Disaster Relief Reimbursements (75-0142-0-1-506):												
Base	0.754	0.765	0.805	0.674	0.866	0.492	0.481	0.293	0.365	0.380	0.260	6.135
Supplemental												
Subtotal,												
Base	0.754	0.765	0.805	0.674	0.866	0.492	0.481	0.293	2.365	2.380	2.260	12.135
Supplemental									600			600
DEPARTMENT OF HOMELAND SECURITY												
Federal Emergency Management Agency, Disaster Relief Fund. With estimates based on historical obligations for major disasters only (70-0702-0-1-453):												
Base	1,453	1,945	744	1,588	1,839	1,486	1,124	1,079	1,242	1,384	2,427	15,726
Supplemental	1,316	5,298	914	2,595	32,946	5,006	3,189	9,063		4,412		70,599
Federal Emergency Management Agency, Disaster Assistance Direct Loan Program, Including Subsidy for the Cost of Loans and Administrative Expenses (70-0705-0-1-455):												
Base	2.105	0.948	1.114	0.560	0.567	0.567	0.569	0.875	0.295	0.295	0.295	6.190
Supplemental						752						752
Subtotal, Department of Homeland Security												
Base	1,455	1,946	745	1,589	1,840	1,487	1,125	1,080	1,242	1,384	2,427	15,735
Supplemental	1,316	8,298	914	2,595	32,946	5,758	3,189	9,063		4,412		77,361
DEPARTMENT OF HOUSING AND URBAN DEVELOPMENT												
Community Planning and Development, Community Development Fund (86-0162-0-1-451):												
Base				0.280	0.391	0.700		0.531	0.053	1		3
Supplemental		3,483			150	16,700		9,423		100		29,856
Public and Indian Housing Program, Tenant-Based Rental Assistance (86-0302-0-1-604):												
Base						10			4	1		14
Supplemental					390				85			475
Public and Indian Housing Program, Project-Based Rental Assistance (86-0303-0-1-604):												
Base												
Supplemental								50	30			80
Subtotal, Department of Housing and Urban Development												
Base				0.280	0.391	1	10	1	4	2		17
Supplemental		3,483			540	16,700		9,473	115	100		30,411
DEPARTMENT OF THE INTERIOR												
Office of Insular Affairs, Assistance to Territories (14-0412-01-808)												
Base	0.500	2		0.580						1.200		4.080
Supplemental												
DEPARTMENT OF TRANSPORTATION												
Federal Highway Administration, Federal-Aid Highways, Derived From the Highway Trust Fund (69-8083-0-7-401):												
Base												
Supplemental	647	440		1,202	662							2,950
Federal Highway Administration, Emergency Relief Program (69-0500-0-1-401):												
Base												
Supplemental						3,382	784		765			4,931

	2001	2002	2003	2004	2005	2006	2007	2008	2009	2010	2011	2001–2011
Subtotal, Department of Transportation												
Base												
Supplemental	647	440		1,202	662	3,582	784		765			7,881
CORPS OF ENGINEERS—CIVIL WORKS												
Investigations (96-3121-0-1-301):												
Base												
Supplemental							8					8
Construction (96-3122-0-1-301):												
Base												
Supplemental							11					11
Flood Control and Coastal Emergencies (96-3125-0-1-301):												
Base												
Supplemental						1,740						1,740
Subtotal, Corps of Engineers—Civil Works												
Base												
Supplemental						1,740	19					1,759
NATIONAL AERONAUTICS AND SPACE ADMINISTRATION												
Space Operations (80-0115-0-1-252):												
Base												
Supplemental					126			30				156
SMALL BUSINESS ADMINISTRATION												
Disaster Loans Program, Including Subsidy for the Cost of Loans and Administrative Expenses (73-1152-0-1-453):												
Base	184	210	191	171	113		115			78	45	1,108
Supplemental	100	75		2,000	929	983		1,053				5,140
TOTAL	4,203	12,454	1,852	7,558	37,157	31,944	5,451	21,365	2,743	6,029	2,475	133,231
Base	1,640	2,158	937	1,781	1,954	1,496	1,250	1,081	1,248	1,468	2,475	17,468
Supplemental	2,563	10,296	914	5,797	35,203	30,448	4,201	20,284	1,495	4,561	—	115,763
TOTAL, 2001–2010												150,756
TOTAL, 2002–2011												129,026

End Notes

[1] Section 102(2) of the Stafford Act defines the term "major disaster" to mean "any natural catastrophe (including any hurricane, tornado, storm, high water, winddriven water, tidal wave, tsunami, earthquake, volcanic eruption, landslide, mudslide, snowstorm, or drought), or, regardless of cause, any fire, flood, or explosion, in any part of the United States, which in the determination of the President causes damage of sufficient severity and magnitude to warrant major disaster assistance under this chapter to supplement the efforts and available resources of States, local governments, and disaster relief organizations in alleviating the damage, loss, hardship, or suffering caused thereby." 42 U.S.C. 5122(2).

[2] Accordingly, OMB's calculation does not reflect total appropriations that were provided to respond to major disasters, but rather only funding that was provided pursuant to appropriations or authorizing language that specifically referenced the Stafford Act.

INDEX

D

E

F

G

H

I

J